Commentary on Article CC-3000 Desig

ASME Boiler and Pressure Vessel Code

Section III, Division 2 Code for Concrete Containments
April 2015

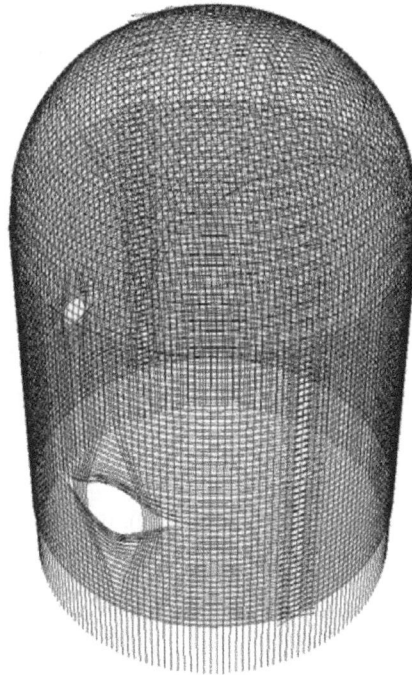

Commentary Task Group

Group Leader
Theodore E. Johnson, TJBG Consulting Inc.

Members
Bill Hovis, AECOM
Ola Jovall, Scanscot Technology AB
Javeed Munshi, Bechtel Corporation
John Stevenson, J.D. Stevenson Consulting, Engineer

Library of Congress Cataloging-in-Publication Data

Names: American Society of Mechanical Engineers. Commentary Task Group. |Johnson, Theodore E., organizer. | Hovis, Bill, researcher. | Jovall, Ola, researcher. | Munshi, Javeed A., researcher. | Stevenson, John D. (John David), 1933-2014, researcher.
Title: Commentary on Article CC-3000 design : ASME boiler and pressure vessel code section III, division 2 code for concrete containments / Commentary
Task Group ; group leader, Theodore E. Johnson ; members, Bill Hovis, Ola Jovall, Javeed Munshi, John Stevenson.
Description: New York : ASME, [2015] | Includes bibliographical references and index.
Identifiers: LCCN 2015040217 | ISBN 9780791861066
Subjects: LCSH: Nuclear reactors--Containment--Standards--United States. |Nuclear pressure vessels--Materials--Standards--United States. |Reinforced concrete construction--Standards--United States. | Code for operation and maintenance of nuclear power plants.
Classification: LCC TK9211 .A44 2015 | DDC 621.48/32--dc23 LC record available at http://lccn.loc.gov/2015040217

DEDICATION

**This Commentary is dedicated to the memory
of John D. Stevenson, PhD**

Dr. Stevenson was a giant in the Nuclear Power Industry in the areas of Structural and Mechanical Engineering and his expertise was recognized throughout the world. He had an incredible history of being in a position of influence from the mid-1960s to October 2014.

John was a member of this Code Committee from the early 1970s and he authored many of the sections of the ASME Section III, Division 2 Code. He was also one of the major contributors to this Commentary.

He served on many committees of the ASCE, ASME, ANS, ACI and AISC charged with the development of standards devoted to design of nuclear power plant and other critical facilities. Dr. Stevenson's relevant experience includes 50 years as a structural-mechanical engineer with particular application to dynamic cyclic impact and impulse load design and analysis.

John was a Consultant to International Atomic Energy Agency Working Groups on the Development of Safety Standards on Extreme Load Design of Nuclear Power Plants and Other Nuclear Facilities; in addition, he was a Division Coordinator for many of the Structural Mechanics in Reactor Technology [SMiRT] Conferences.

His consulting was to Engineering Companies, Power Utilities, and Nuclear Power Agencies worldwide.

Dr. Stevenson published more than 100 papers and gave invited talks at large corporations and national labs, including Oakridge, Lawrence Livermore, Argonne, and Los Alamos, and International Conferences and Agencies, including SMiRT conferences and the IAEA.

In 1981, he published as a Chair Editor the reference book for the nuclear industry on "Structural Analysis and Design for Nuclear Facilities" by ASCE. In 1991, Dr. Stevenson received several prestigious awards of ASCE and ASME as a national recognition of his professional accomplishments.

ACKNOWLEDGMENTS

Gunnar Harstead, Harstead Consulting
Michael Thumm, AECOM

Technical Editors:

Ying Fang, AECOM
Colin Manning, AECOM
Jessica Stratton, AECOM
Jordan Supler, AECOM
Michael Tedeschi, AECOM

CONTENTS

The Appendices herein either contain the full material, excerpts or a summary depending on copyright restrictions. These Appendices contain what are considered as very informative material.

REFERENCES

1. Rajgopaian, K. J., and P. N. Ferguson, "Exploratory Shear Tests Emphasizing Percentage of Longitudinal Steel," ACI Journal, pp. 634–638, August 1968.

2. ACI-ASCE Committee 326, "Shear and Diagonal Tension," ACI Proceedings, Vol. 59, No. 1, pp. 1-30, 1962; No. 2, pp. 277–334, 1962; No. 3, pp. 352-396, 1962.

3. MacGregor, J. G., and J. M. Hanson, "Proposed Changes In Shear Provisions for Reinforced and Prestressed Concrete Beams," ACI Proceedings Vol. 66, No. 4, pp. 275–288, 1969.

4. U.S. Nuclear Regulatory Commission, Crystal River Unit 3, Containment Delamination Update by Progress Energy, November 2009.

5. ACI-ASCE Committee 423, "Tentative Recommendations for Prestressed Concrete," ACI Proceedings, Vol. 54, No. 7, pp. 545–578, 1958.

6. ACI Committee 435, "Deflections of Prestressed Concrete Members," ACI Proceedings, Vol. 60, No. 12, pp. 1697–1728, 1963.

7. Troxel, C. A., "Tests and Design of Bombproof Structures of Reinforced Concrete," Navy Department, U.S. Government Printing Office, Washington D.C., 1941.

8. Dunham, C. W., "The Theory and Practice of Reinforced Concrete," Second Edition, McGraw-Hill Inc., 1944, p. 298.

9. Topical Report BC-TOP-7, Full Scall Buttress Test for Prestressed Nuclear Containment Structures, T. Johnson, R. Marsh, September 1972, Bechtel Corporation.

10. Topical Report BC-TOP-8, Tendon End Anchor Reinforment Test, H. Franklin, T. Johnson, September 1972, Bechtel Corporation.

11. U.S. Nuclear Regulatory Commission, Standard Review Plan for the Review of Safety Analysis Reports for Nuclear Power Plants (NUREG-0800).

12. U.S. Nuclear Regulatory Commission, NUREG/CR-6906, Containment Integrity Research at Sandia National Laboratories, September 12, 2006.

13. U.S. Nuclear Regulatory Commission, Overpressurization Test of a 1:4-Scale Prestressed Concrete Containment Vessel Model (NUREG/CR-6810), Sandia National Laboratories, March 2003.

14. Moreadith, F., and R. Pages, "Delaminated Prestressed Concrete Dome: Investigation and Repair" ASCE, 1983.

15. Basu, P. C., Gupchup, V., and Bishnoi, L. "Containment Dome Delamination, SMiRT 16, Washington DC, August 2001.

INTRODUCTION

This commentary discusses some of the considerations of the joint ACI-ASME Committee in developing the provisions of ACI Standard 359 and ASME B&PVC Section III, Division 2, Subsection CC, Article CC-3000 in the 2013 version of the code. Emphasis is given to the explanation of provisions that may be unfamiliar to code users. Comments on specific provisions are made under the corresponding paragraph numbers of the code. The figures and appendices referred to in this commentary occur only in the commentary so that their numbering has no parallel in the code.

Because the code is written and intended for use as a legal document, it does not present background details or suggestions for carrying out its requirements or intent. It is the intent of this commentary to at least partially fill this need. This commentary also directs attention to other documents that provide suggestions for carrying out the requirements and intent of the code. However, neither those documents nor this commentary are to be considered as a part of the code.

The commentary is not intended, in general, to provide a historical background concerning the development of CC-3000, nor is it intended to provide a detailed résumé of the studies and research data reviewed by the committee in formulating its provisions. CC-3000 is based on, or extrapolated from, ACI 318-77 to a large extent, but an effort is presently underway to update it based on recent ACI Codes, where applicable.

References to some of the research data are provided for those who wish to study the background material in depth.

This code is meant to be used as part of a legally adopted code associated with the construction of nuclear reactor containments and, as such, must differ in form and substance from documents that provide recommended practice, complete design procedures, or design aids.

The code is intended to cover current types of concrete containments. Requirements more stringent than the code provisions may be desirable for unusual construction. The code provides only the minimum requirements necessary to provide for public health and safety. For any containment, the Owner or the Designer may require the quality of materials and construction to be higher than the minimum requirements necessary to protect the public as stated in the code. However, lower standards are not permitted.

The code has no legal status unless it is adopted by governmental bodies having the legal power to regulate containment design and construction. Where the code has not been adopted, it may serve as a reference to good design practice even though it has no legal status. The code provides a means of establishing minimum standards for acceptance of designs and construction by a legally appointed Authorized Nuclear Inspector.

It should be noted that nuclear power plant containments are assumed to always be capable of performing their design function. They are unique among engineered safeguards in that they are assumed to be effective in protecting against the unmitigated release. As a result, they require a very robust design where design basis accident loads use load factors that can approach those for normal operation and design.

Since a significant amount of reference material was published 30 to 40 years ago and is difficult to find, some of it has been included in the Appendices of this commentary.

BACKGROUND

Illustrated in Appendix A are reports and figures that show some of the various Containment configurations that have been or are being designed and built either in accordance with this code or using it as a guide document. This information is very informative and should be reviewed in detail. [A-1] Advances in the Analysis and Design of Concrete Structures, Metal Containments and Liner Plate for Extreme Loads, [A-2] History of the Development of United States Prestressed Concrete Containments (PCC) and [A-3] Figures in this Appendix illustrate various configurations of Concrete Containments. These include both past and those presently being considered for use in the USA.

ARTICLE CC-3000 DESIGN

CC-3100 GENERAL DESIGN

The code presents criteria for the design of a safe, reliable nuclear power plant containment. These criteria have been extrapolated from conventional codes and experience that has been built up through involvement with actual containment designs and testing programs.

1. GENERAL OVERVIEW OF DESIGN AND ANALYSIS TECHNIQUES

When the design of reinforced and fully prestressed containments started in about 1966, analytical procedures were generally limited to those associated with elastic shell theory. Alternatively there were only limited structural analysis computer programs available. These finite element computer programs could solve axisymmetric structures with axisymmetric loads in uncracked concrete. As time went by, there were programs developed which could handle concrete cracking and model the reinforcing steel and perform nonlinear analysis.

The present ASME Section III, Division 2 Code allows the use of elastic structural analysis. With the code approach it is very desirable to have a post-processor that takes the elastic computed section forces and moments and then does the design in accordance with the Code allowables. The code allows nonlinear concrete compression stress distributions and allows the reduction in thermal effects due to concrete cracking. Also, in some cases reinforcing strains may exceed yield where the forces and moments considered are deformation-limited.

2. CONCRETE STRUCTURAL DESIGN AND ANALYSIS

The information provided below pertains to the type of analysis and design that would satisfy code requirements.

2.1 General Shell

2.1.1 Manual Calculations It is important to have manual calculations done as an overall check on the computer analysis. Appendix B illustrates the treatment of a cylinder to base mat junction. Appendix D illustrates manual calculations for both a Prestressed and a Reinforced Concrete Containment.

2.1.2 Finite Element Analysis There are many good finite element programs available. It is important that the program documentation meets industry and NRC requirements. It is extremely important

to have a post-processor that takes the section forces and moments together with thermal effects and computes resultant stresses, strains, etc. Then these computed values should be compared with Code allowables. This entire process should be automated with auditable results. The overall analysis is based on elastic analysis. However, the analysis of sections includes concrete cracking and the deformation-limited treatment of thermal bulk gradients and liner plate thermal expansion. Also, when computing stresses, the net concrete area should be used. That is, the tendon duct [sheathing] area shall be deducted for unbonded tendons.

Some typical finite element mesh layouts are included in Appendix B. Two papers on post-processors are also included in this Appendix.

It should be noted that while this code uses load factors to develop input loads, the resultant acceptance criteria is based on stresses or strains associated with Allowable Stress Design, ASD, and not Strength Design, SD, as typically used in concrete building design.

2.2 Discontinuities and Base Mat

These areas typically use a smaller mesh near the discontinuity. When analyzing the base mat it is important to make sure the contact of the lower surface has no tension resistance from the foundation material under the containment. Treating the cylinder base and the base slab junction in an elastic analysis is conservative for moments and shear forces since concrete cracking would reduce the calculated moments and radial shear, but this may increase the computed membrane tensile loads in the shell and base mat junction.

2.3 Equipment Hatch and Access Locks

These areas also typically use a smaller mesh. A simple rule is to replace the material removed by a combination of concrete and reinforcing steel. Typically only additional reinforcing is used for the smaller penetrations

CC-3110 Concrete Containment

(b) The purpose of a containment is to isolate the possible consequences of the release of the energy contained in the reactor coolant system, a nuclear incident, from the surrounding environment. To assure that this requirement is clearly defined, the Design Specification, as required in NCA-3250, must clearly identify all functional requirements and design parameters of the containment.

(d) In CC-3110 (d) (1) and (2), the Code recognizes that ductile behavior exists in a properly designed containment structure. There are three stages or modes, of behavior. See Figure CC-3110-1. These include:

(1) Elastic – Where all reinforcing bars are at strains less than yield strain.
(2) Partial Yield – Where, in some reinforcing bars, yielding has occurred with resulting reduced stiffness.
(3) General Yield – Where sufficient reinforcing bars have yielded to allow deformation without additional load.

In CC-3110 (d) (1), the Code allows behavior in Stage 2 for primary loads which are not self-limiting in nature. This is permissible for several reasons:

(a) The inherent ductility of commonly used reinforcing bar steel with specified minimum yield and ultimate strength equal to or less than 60 and 90 ksi, respectively, is specified. Use of mechanical or welded reinforcing splices rather than lapped splices in biaxial tension fields is required.
(b) Reliable reinforcing bar couplers.
(c) A large number of bars, which minimizes the effects of any possible flaw in an individual bar.
(d) The dominance of membrane tension in most areas of the structure in a non-post-tensioned containment, which precludes brittle failure due to excessive concrete compression strain.
(e) The progressive nature of reinforcement yielding inherent in the multilayer reinforcing patterns generally present, which results in gradual reduction in stiffness as progressive layers yield, and does not result in the sudden occurrence of Stage 3.

An example of a case where Stage 2 behavior is acceptable is yielding of diagonal membrane shear reinforcement in a four-way reinforcing system which is stressed in tension by membrane forces as well as earthquake membrane in-plane shear. This is acceptable behavior since only one of the diagonal bars is stressed in tension by the meridian and hoop forces as well as by the in-plane shear force. Typically, the orthogonal reinforcement adjacent to the diagonal reinforcement is stressed to levels significantly lower than $0.9 f_y$. Yielding of the diagonal reinforcement signifies the onset of partial yield. Yielding must develop in another reinforcement group (hoop or meridional or the other inclined layer) to produce a general yield condition. Where the inclined reinforcement is proportioned to resist all the membrane shear force in excess of that taken by the concrete, as required by this Code, limited reinforcement yielding is acceptable in one layer of diagonal reinforcement bars.

Other examples of Stage 2 behavior are found in CC-3422.1-C-(1).

In CC-3110 (d) (2), the Code allows limited behavior in Stage 3 when secondary effects are considered. These are primarily volume change effects and include thermal effects such as the increase in liner temperature due to accident thermal transients inside the containment and thermal gradients through the containment wall. These are typified by their self-relieving nature. Since these loads do not affect the overall load carrying capability of the structure, behavior in Stage 3 is acceptable, provided that, in load combinations with primary and secondary loads, consideration of the primary loads alone does not produce behavior in Stage 3.

In addition, one must ensure that significant effects of the yielding permitted above are considered. For example, additional concrete cracking and redistribution of loads must be considered. Additional cautions to control the effect of reinforcement yielding are discussed in CC-3422.1(d).

The tensile strain in reinforcing steel resisting local section shear forces is not permitted to exceed yield.

Structural members designed to resist impulse loads and impact effects are permitted to exceed yield strain and displacement values. Ductility limits are given by the code for use in evaluating the structure. Design criteria are provided in CC-3900.

CC-3120 Metallic Liner
CC-3121 General

The purpose of the liner and liner anchorage system is to accommodate all design loads and containment deformations without jeopardizing the leak tight integrity of the structure. For this purpose the code permits only the use of a suitably ductile metallic liner.

The liner is not permitted to be used as a strength element in designing the containment to resist loads such as accident pressure and earthquake. However, under test conditions, the liner is in tension and should be considered to predict test results.

During load combinations including thermal effects, where the liner interacts with the concrete shell and transmits thermal loads to the shell, or expansion of the containment due to pressure loads it is necessary to include the interaction of the liner and the concrete to properly evaluate strains in the liner and concrete. For consideration of primary loads alone, however, the liner cannot be considered a strength element.

The liner has been used as a structural element to resist construction loads such as concrete placement loads. In addition the liner has also been used as a structural element to resist small loads from items such as electrical cable tray and conduit supports, ladders, etc.

CC-3122 Liner

Liner stress at or above the buckling stress of a constrained plate does not describe an unstable condition, since, through the liner anchorage system, the liner is forced to act compatibly with the concrete. Although this stressed state is reached, the liner system is still generally constrained by the anchorage

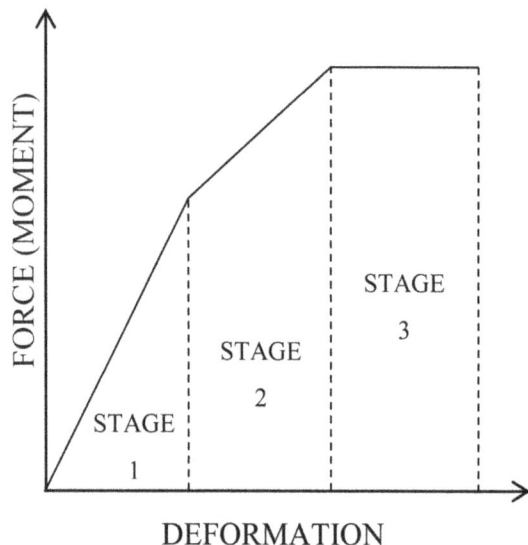

FIGURE CC-3110-1 STAGES OF REINFORCEMENT BEHAVIOR.

system, bears against the concrete and is still capable of performing its function as a leak-tight membrane. As long as the liner is able to perform this function within the deformation limits and strain in CC-3700, the stability stress level does not have to be considered. Care must be taken, however, to ensure that the integrity of the anchorage system is maintained in the deformed state.

In order to maintain good liner ductility, all liner plate seams should use full penetration butt welds. Joints should not be made with fillet welds.

CC-3123 Liner Anchors

Allowable liner anchor behavior is based on ultimate deformation, which is generally determined by a testing program. The liner anchorage system is not stress-limited since its primary function is to anchor the liner to the concrete and limit liner strain so that liner integrity is maintained. Inward displacement under compression loads from effects such as thermal will most likely occur for liner areas that are either flat or have initial inward curvature. This condition is acceptable. However, the anchorage system must be designed for the resulting loads.

This can simply be satisfied by considering a one way plate strip with no resisting elements on the side opposite of the applied load, then showing that the anchor has sufficient strength and displacement capability to accommodate the loads.

CC-3136 Stresses and Forces

Two important aspects of design rules are the defined loadings and the acceptable limits and modes of behavior of the structural system in response to these loadings.

The first aspect is covered by load combinations and load factors, while the second is covered by limitations on usable stress and strain. The limitations of this code are generally related to stress and strain and not strength as in other ACI Strength, SD, Codes. However, a limit on stress and strain achieves essentially the same strength levels as indicated in Tables CC-3421-1 and CC-3431-1 while also indirectly achieving predictability of the serviceability requirements.

In establishing the limits, both the behavior of the two main structural elements and the system as a whole are considered. The two main elements of a containment are concrete and steel. Concrete behavior is essentially non-ductile and it is therefore appropriate to specify limits on stress in the elastic range. Steel, on the other hand, is ductile and it is therefore appropriate to specify both stress and strain limits. By specifying appropriate limits on the elements, acceptable behavior of the composite system can also be assured.

Levels of acceptance limits are a function of modes of behavior. To be consistent with ASME Codes, two modes of behavior and associated stress limits have been established i.e., primary and secondary. Primary mode behavior is the ability of internal forces to equilibrate applied loads. Secondary resistance is exhibited when, (1) internal forces are not required to balance external forces, or (2) the external loads are self-limited.

Examples of the first type of secondary behavior are the redundancy exhibited at the juncture of the containment wall and base slab (meridional moment and hoop membrane forces), and redistribution of forces in a four-way reinforcing system for tangential shear. An example of the second type of secondary behavior is internal forces resulting from volume changes such as thermal and shrinkage effects.

Bending at regions of discontinuity is listed as secondary since it is recognized that the internal force is self-limiting when proper attention is given to the primary load carrying system. However, it is not permissible to design for Stage 3 (See CC-3110(d)) behavior in these regions. The design shall be based on requirements for primary bending moment in CC-3422.1(c) (1).

Where there is a redundancy of the internal force system, the primary, or load equilibrating force system, must be justified in the Design Report. For example, the moment at the base of the cylinder (presently designed for Stage 2 behavior) is secondary because the designer generally chooses to provide hoop reinforcement to resist all hoop forces. The designer could choose to eliminate some hoop reinforcement and carry load in meridional bending in which case this would become a primary force.

Primary internal forces are limited to Stage 1 and 2 behavior and secondary internal forces are limited to Stage 2 behavior for redundancy and Stage 3 behavior for self-limiting volume change effects (See CC-3110(d)).

Because of the nature of secondary stress behavior, it is appropriate to allow higher limits which allow and assure elastic shakedown. For steel this is accomplished by allowing higher stress and strain limits. For concrete this is accomplished by allowing higher stress limits. Also, for concrete, higher limits are established for bending stress than for membrane stress, as bending stress only utilizes a small portion of the cross-section and redistribution of stresses so a more uniform stress block is possible where there is sufficient ductility to accommodate deformation limited loads in the absence of significant cyclic loads.

CC-3200 LOAD CRITERIA

This section provides loads and load combinations for design of containments. These include service loads and factored loads. Service loads include normal loads, construction loads, and test loads. Factored loads include severe environmental loads, extreme environmental loads, and abnormal loads.

Computations for prestressing loads in prestressed containments should consider the tendon tensioning sequence, the prestress load immediately after transfer of prestress, including prestress loss, and the variation of prestress along the tendon due to friction.

Temperature effects include those induced by the restrained thermal expansion of the containment and the steel liner and temperature gradients. Transient temperature conditions during startup and shutdown should be considered. The combination of concrete internal and ambient temperatures that produces the maximum effects should be used. The self-relieving nature of the liner plate expansion load and thermal gradients may be considered.

CC-3221 Service Loads

The loads listed in the service load category are expected to occur during plant construction, testing, operation, and shutdown during the design life.

CC-3222 Factored Loads

Factored loads are loads that are unlikely to occur during the plant lifetime. However, they are included as a design basis so that public safety is assured in the event that one or more of these loads does occur, and so the plant can be maintained in a safe shutdown condition. Factored loads should not be confused with load factors which are applied to individual loads in order to obtain the desired level of conservatism in design.

It should be noted that the unmitigated release of radioactivity is usually considered as part of the safety analysis for nuclear facilities. Nuclear reactor containments are unique in Safety Analysis in that their failure to mitigate the release of radioactive inventories is not assumed as a design basis. For this reason concrete reactor containments are robustly designed such that design basis accident loads (i.e., design basis pressure) are treated as normal design loads which employ a load factor of 1.5.

CC-3222.1 Severe Environmental Loads

The severe environmental loads are the design basis loads which, while not expected, could be developed during the operating life of the plant. It is generally assumed the plant would be operable after such an event (i.e., typically a mean 100 to 200 year return period for wind, precipitation (snow or rain) and flooding). It should also be noted these severe loads have load factors greater than one which results in design basis severe loads with return periods significantly larger than 100 years.

In cases where an Operational Basis Earthquake (E_o) peak ground acceleration (pga) exceeds one-third of the Safe Shutdown Earthquake (E_{ss}) pga, it may also be defined as a Severe Environmental design basis load.

CC-3222.2 Extreme Environmental Loads
The extreme environmental loads are the maximum loads such as the Safe Shutdown Earthquake and tornado, for which the plant must be maintained in a safe shutdown condition.

CC-3222.3 Abnormal Loads
Abnormal loads result from design bases accidents. When combined with Design Basis Accident procedures, thermal effects shall be calculated using upper limit values based on the design accident pressure "P_a" and not the design accident condition "$1.5P_a$." The peak design pressure may be combined with the concurrent thermal effect and the peak design temperature with concurrent pressure if justified by a time-history analysis.

The dynamic effects of high energy pipe rupture capable of impacting the containment must also be considered in the containment design.

CC-3222.4 Jurisdictional Boundaries

It should also be recognized that there is a need to define Jurisdictional Boundaries between this concrete containment code and other construction codes (i.e., ACI 349) when the base mat and other parts of the containment are formed integrally with adjacent structures.

The relative stiffness of the foundation media of the containment affects how far the containment pressurization loads will extend into a common base mat for the containment and a nuclear island outside the containment geometric boundary. For a base mat with a constant thickness, based on a parametrical study carried out by Scanscot Technology AB (Lund, Sweden), the distance, as a function of foundation media support stiffness, for which the influence has been reduced significantly could be estimated as follows:

(1) Three times the base mat thickness measured from the outer boundary of the containment structure for foundation shear wave foundation media that has a shear wave velocity equal to or less than 1000 ft/sec (300 m/sec).

(2) Two times the base mat thickness measured from the outer boundary of the containment structure for foundation shear wave foundation media that has a shear wave velocity equal to or less than 2500 ft/sec (750 m/sec).

(3) One times the base mat thickness measured from the other boundary of the containment structure for foundation shear wave foundation media that has a shear wave velocity greater than 2500 ft/sec (750 m/sec).

For other parts integrally formed with the containment pressure boundary, the effect the integral structures have on the resistance of the containment pressurization load have to be judged on a case-by-case basis.

The Jurisdictional Boundaries should at least be placed one rebar development length outside the outer part of the containment pressure boundary; however, it is recommended to also consider how important adjoining structures are in resisting the containment pressurization load when determining the Jurisdictional Boundaries.

It shall be noted that the Jurisdictional Boundaries between this concrete containment code and other construction codes determine the requirements for all parts of the construction. Independent of the position of the Jurisdictional Boundaries, it is recommended that the Design of integral parts that significantly contribute to the resistance of the containment pressurization load, in addition to the construction code in charge (if other than this concrete containment code), also is designed according to CC-3000.

CC-3230 Load Combinations

The load factors in the service category (Table CC-3230-1) are unity since the design allowables are well below yield and are similar to what is commonly referred to as "working or allowable stress design." Since these loads are assumed to occur during the operating life of the plant, the predicted overall structural response is kept well below yield.

The load factors in the factored category are unity or greater, depending on the probability of load occurrence, confidence in load magnitude and required safety margins associated with a robust design.

The combinations and factors given in the Severe Environmental Category are similar but not necessarily the same as contained in ASCE 7-05 Standard and generally assume a 100 year mean return period event. The E_o load is considered a design basis load when its pga is greater than one-third the E_{ss} pga. There are a number of other severe natural hazard loads such as precipitation (snow, rain, ice) and flooding which are defined in ASCE 7-05 that are assumed to be enveloped by W or W' and E_o or E_{ss} so they are not specifically considered in Table 3230-1. This assumption is currently under review and may result in the specific inclusion of precipitation and flood loads in the future. In this assessment it is assumed that the ASCE 7-05 Importance Factor is equivalent to increasing return periods from 50 to 100 years.

The unity load factors given for the Extreme Environmental Category are justified since these are considered extreme upper limit design basis loadings and the allowables for concrete are based on specified minimum fractions of concrete ultimate stress and yield in reinforcing steel and not ultimate capacity.

The load factors given in the Abnormal/Severe Environmental Category are considered justifiable based on the low probability of occurrence of these loads. The factor of 1.25 on pressure, operational basis earthquake, and design wind is considered a sufficient margin.

The load factors given in the Abnormal/Severe Environmental Categories are considered justifiable based on the low probability of occurrence of these loads.

The unit load factors given in the Abnormal/Extreme Environmental Category are sufficient since upper limit loads are being combined and the structural capacity to resist these loads is based on yield and not ultimate strength.

The U.S. NRC in its Regulatory Guide 1.136 "Design Limits Loading Conditions, Materials, Construction and Testing of Concrete

Containment" Rev. 3, U.S. Nuclear Regulatory Commission, March 2001, includes a loading combination which assumes loss of coolant and some level of reactor core melt resulting in the release of hydrogen associated there with which can be characterized as a Beyond Design Basis Loading which does not assume a concurrent seismic load.

The U.S. NRC has also identified another Beyond Design Basis Earthquake Load which is equal to 1.67 times the E_{ss} load but also does not appear to be taken concurrent with a P_a, R_a, or T_a loadings.

Currently the ASME B&PCV code does not address Beyond Design Basis Loads on their associated acceptance criteria.

CC-3300 CONTAINMENT DESIGN ANALYSIS PROCEDURES

The Code does not prescribe specific analytical procedures, but leaves the choice of analytical techniques to the Designer. Appendix D has some sample design calculations for both Prestressed [D-1] and Reinforced [D-2] Containments.

In considering the techniques available for the analysis of a containment, the Designer must consider the various elements that constitute this component of the nuclear power plant. Consequently, Subarticle CC-3300 offers comments on the three major items that make up the containment structure. These typically are the shell, the more conventional elements, including the base mat, frames, and slab assemblies, and the penetrations and openings.

With regard to the analysis of the general shell, the Code recognizes, for overall pressure load, linear elastic analysis using shell theory to include base slab-wall connection, wall-dome connections, penetrations, ring girder, polar crane supports or other discontinuities where significant bending of the shell occurs. Techniques involving various methods of discretization such as finite elements or difference are acceptable.

Whatever recognized method of structural analysis is used, the results must be reviewed for consistency, reasonableness of the response, and agreement with the initial boundary conditions specified, as well as for verification of the equilibrium of the structural system.

The analyses of elements such as the base mat can be those used in normal base mat conventional design practice. Consideration must be given to the range of support properties and influence of superstructure interaction that will exist as a function of the design loading on the containment. If the mat is analyzed as a separate element of the containment, compatibility checks must be made at the junctures with other elements of the containment. It is now common to include the base mat as a part of a larger FEM model for the nuclear island structures.

For large openings such as personnel lock and equipment hatch, the results of analysis treating the containment as a shell of revolution are applicable only to sections of the shell sufficiently far from the opening. The state of stress and deformation near such a large opening (i.e. greater than twice the wall thickness) must therefore be investigated by considering a suitable localized or free body of that region of the shell and using forces or deformations obtained from the shell of revolution analysis at the boundary of the localized free body representation of the large opening. For smaller opening areas, replacement of steel reinforcement in the local concrete areas around the opening should be applied.

Model testing and analytical correlations such as the tests sponsored by the NRC at the Sandia National Laboratory, does offer support to various analytical simplifications and other concepts that may be used in analysis, see U.S. Nuclear Regulatory Commission NUREG/CR-6906 (Reference 12). Model testing programs should clearly define how they represent the prototype, what the limitations and the test objectives are. The results obtained should be provided along with the decision logic that will be formed as a result of the various facts determined from the testing.

The code requires that temperature effects be considered in the containment design.

ACI 349-06 has information on the treatment of thermal effects and can be used for guidance; however, the requirements of ASME Section III, Division 2 must be satisfied.

CC-3400 CONCRETE CONTAINMENT STRUCTURE DESIGN ALLOWABLES

CC-3410 General

The containment structure for which this Code has been developed differs from the type of structure covered by the ACI 318 and ACI 349 Codes. This Code requires both factored load and service load design (Allowable Stress Design, ASD), using limiting stress values, whereas ACI 318 and ACI 349 use the strength method, where only section capacities in terms of forces and moments are defined, or the alternate limiting stress method. The stresses for factored load design in this Code are established so that general yielding of cross sections will not be reached except as discussed in CC-3110. The limiting stresses for service load conditions are established to maintain overall elastic behavior and to ensure that the structure, including both structural and leak-tight integrity, will not deteriorate in service. The appropriate capacity reduction factor is already included in the specified design values. It should be understood that the containment structure, unlike conventional structures, performs a pressure retaining low leakage function. For this reason, limiting stress behavior criteria, which controls deformation of the structure as well as strength capacity, has been used for all load categories.

CC-3421 Concrete

Conforming to stress allowables for concrete given in 3421.1. Figure CC-3421-1, below, gives a recommended concrete stress-strain curve for use in developing a load-moment interaction

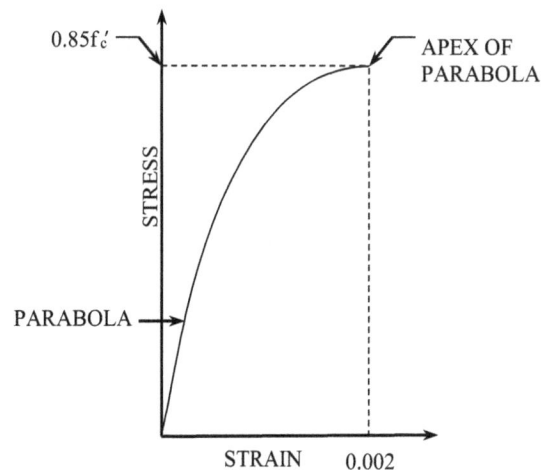

FIGURE CC-3421-1 RECOMMENDED CONCRETE STRESS-STRAIN CURVE.

diagram to determine the adequacy of a cross section subject to membrane load and moment. However, as stated in CC-3511.1(e), other curves may be used.

CC-3421.1 Compression The primary membrane compression value of $0.6\ f'_c$ is considered to be a reasonable approximation of the ultimate compression strengths of large, long concrete members. The primary membrane plus bending compression value of $0.75\ f'_c$ is considered a reasonable limit since it is lower than the $0.85\ f'_c$ used in the concrete stress-strain curve, as shown in Figure CC-3421-1. Higher allowable stresses are permitted if self-limiting secondary stresses are considered in addition to the primary stresses.

CC-3421.4 Radial Shear Radial shear primarily occurs at structural discontinuities such as cylinder to base slab intersections, cylinder to shallow dome junctures and in the concrete surrounding large openings.

CC-3421.4.1 Reinforced Concrete Equations (1a) through (1c) are the basic equations for shear strength of members. These equations are similar to Equation (11-6) of ACI 318-77 except that Equation (11-6) was modified to reflect test results (Reference 1). These test results indicated that Equation (11-6) should be modified for low values of ρ.

Equations (2) and (4), for members subjected to axial compression in addition to shear and flexure, are derived in the ACI-ASCE Committee 326 report (Reference 2).

Another design provision, Equation (4), is included for the case of axial tension existing with shear and flexure. These equations were discussed and comparisons were made with test data in (Reference 3).

CC-3421.4.2 Prestressed Concrete
1.0 General and background
Sometime in the mid to late 1960's, Professor Alan Mattock developed a technique to design for radial [through thickness] shear in prestressed concrete containment shells. Professor Mattock was on the faculty at the University of Washington and had done significant research on shear in prestressed concrete.

Radial shear in shallow dome containments primarily occurs at structural discontinuities such as, at the cylinder to base slab intersection and at the ring girder.

The Mattock technique was used to check radial shear capacity and to determine if additional reinforcing steel was required.

The Mattock technique was incorporated into the ASME Section III, Division 2 Code in the early 1970's. However, at a later date, it was somewhat simplified to make the application more straight forward. The provisions have remained the same for many years. Appendix C has a design example of radial shear considerations, [C-1] Radial Shear Tie Example and a [C-2] Radial Tension Tie Example.

Equation 6 is based on the approach in ACI 318-77 for calculating v_{cw}, except ACI 318 allows $4\sqrt{f'_c}$ principal tensile stress and the code only allows $3.5\sqrt{f'_c}$.

Equation 7 is similar to Equation (11-11) in Section 11.4.2.1 of ACI 318-77 except for the following:

(a) A constant value of 0.6 has been replaced by K in this code to account for the effect of the reinforcing steel.
(b) The code requires that the loads be considered in their chronological loading order. Or, in case the order is not known, rules are given to evaluate the most conservative loading sequence.

In Appendix C, an example illustrates the design techniques given in CC-3421.4.2. The loading values used in the example are summarized in Table DE2-1 for the junction of a containment cylinder to the base slab. In the example, the moments and shears caused by vertical membrane forces have been neglected.

CC-3421.6 – Peripheral Shear The procedure shown is an averaging process required because the membrane stresses, and thus the allowable concrete shear stresses, are not necessarily equal in the meridional and hoop directions. The expressions for v_{ch} and v_{cm} were derived using Mohr's circle and define the magnitude of shearing stress which, in combination with either hoop or meridional membrane stress, will produce a principal tensile stress of $4\sqrt{f'_c}$.

When f_h or f_m are tension and greater in magnitude than $4\sqrt{f'_c}$ the equations are not applicable. The specific minimum value of v_{ch} and v_{cm} shall be applicable for the case where f_h or f_m are tension and greater than $4\sqrt{f'_c}$. The calculation of v_c as a "weighted average" of v_{ch} and v_{cm} may be illustrated by considering a rectangular loaded area with the meridional dimension equal to twice the dimension in the hoop direction. The term v_{ch} is the permissible shear stress on a surface parallel to the hoop direction, and v_{cm} is the permissible shear stress on a surface parallel to the meridional direction. Thus, for this example, $v_c = (2v_{cm} + v_{ch})/3$ is the allowable shear stress resisted by the concrete.

CC-3421.9 Bearing Allowable bearing stresses may be increased when the loaded area is confined by beams other than a larger supporting surface, provided that the confinement is capable of resisting the splitting stresses developed by the loaded area. An example of the above confinement is a steel plate on the outside of the loaded area often found at containment penetrations.

CC-3422 Reinforcing Steel
CC-3422.1 Tension
(a) An upper limit of 60,000psi (400MPa) is placed on the yield strength, f_y, of nonprestressed reinforcement. This is due to the decrease in ductility typically exhibited by higher-strength reinforcement above 60,000psi (400MPa).
(b) For load resisting purposes, the allowable stress in reinforcement is limited to $0.9\ f_y$. This does not imply that reinforcement strains cannot exceed yield strain as allowed in other sections of the Code. However, when considering the properties of the stress-strain characteristics for reinforcement, to calculate the load-resisting capability of a section, an elasto-plastic stress-strain curve with a maximum stress of $0.9\ f_y$ must be considered.
(c) Behavior in Stage 2 as discussed in Code Commentary CC-3110 (d) is permitted for Primary Bending Moment, tension rebar diagonal and adjacent to large openings, provided the specified requirements are satisfied. Presently, the code considers bending near discontinuities in the same manner as Primary Bending Moment even though it is classified as secondary by Table CC-3136.5-1.

When tension and compression diagonal rebar are tied at their intersections to avoid postulated buckling of the yielded tension or compression diagonal rebar during load cycling, the following equation may be used:

$$F = \frac{f_s A_s s}{2R} \text{ (See Figure CC-3422-1)}$$

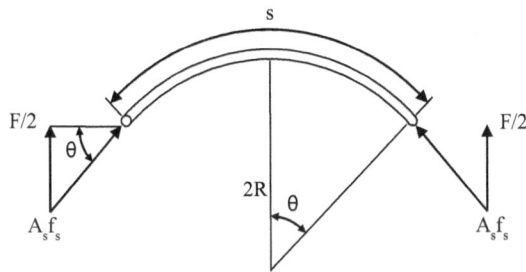

FIGURE CC-3422-1   DESIGN OF DIAGONAL REBAR TIE.

Where

    F = The required connector or tie strength

    A_s = The area of the compression diagonal

    f_s = The maximum compressive stress of the diagonal reinforcement. Use the specified tensile yield strength of reinforcement (f) unless it can be shown that this stress level will not be reached during load cycling

    s = The tie spacing (not to exceed 16 bar diameters)

    R = Containment radius to inclined bars

  (d) To control concrete cracking and ensure limited liner strains, the net tensile strain in the rebar is limited to $2\varepsilon_y$. This further ensures that excessive structural distortions that could affect the serviceability of the containment will not be permitted by the Code. The concept of net tensile strain is introduced to exclude non-stress-related thermal strain from the limitation of $2\varepsilon_y$.

CC-3423 Tendon System Stresses

The intent of this Subparagraph is to permit an increase in tendon stress to be utilized when the applied loads cause tensile strain in the tendons. For unbonded tendons, the computed stress must reflect strain conditions along the tendon. It is obvious that when a design condition, such as $1.5P_a$, is considered, the concrete will crack if the effective prestress level is about $1.25P_a$, and the tendon stress level will increase. However, if cracking results from a design condition such as $1.25P_a + 1.25E_o$, then the amount of cracking will vary throughout the structure. This must be considered in determining the amount of tendon stress increase after cracking.

CC-3430 Allowable Stresses For Service Loads

CC-3431 Concrete Stresses

CC-3431.1 Compression The allowable stresses for service compression membrane and bending loads are the same as those specified in Section B.3.1 of ACI 318-77 ($0.45\ f_c'$). The limiting primary membrane compressive stress is $0.30\ f_c'$.

Higher primary compressive stresses have been permitted for the short-term loadings due to W or E_o and the loads at initial prestress. No increase in the primary membrane plus bending stress allowables has been permitted for such loading conditions. The limiting compression stress for membrane plus bending has been increased by one-third when secondary effects are combined with other loads.

$$\frac{F/2}{A_s F_s} = \sin\theta$$

The radius of a curve's 45-degree inclined bar is 2 times the radius from the center of the containment to the plain of reinforcing.

$$\sin\theta = \frac{s/2}{2R}$$

(FOR LARGE RADIUS)

$$\frac{F/2}{A_s f_s} = \frac{s/2}{2R}$$

$$F = \frac{A_s f_s s}{2R}$$

CC-3431.2 Tension This provision is consistent with Sections B.5.3 and 10.2.5 of ACI 318-77, with one exception. The tensile stresses in the concrete permitted by Section 18.4 of ACI 318-77 are not permitted in this Code.

CC-3431.3 Shear, Torsion, and Bearing This subparagraph is essentially the same as Appendix B of ACI 318-77. A one-third stress increase has been permitted for the temporary loads from prestressing and the test condition and when secondary effects are combined with other loads. No increase in the primary stress allowables is permitted in load combinations that include either W or E_o. The allowable bearing stresses are those specified in Appendix B of ACI 318-77 and are considered to be adequately conservative for prestressing applications in containments.

CC-3432 Reinforcing Steel Stresses and Strains

A one-third stress increase has been permitted for the temporary loads from prestressing and the test condition and when secondary effects are combined with other loads. No increase in the primary stress allowables is permitted in load combinations that include either W or E_o.

CC-3433 Tendon System Stresses

  (a) The maximum stress in the tendon during stressing shall not exceed $0.8f_{pu}$. This will introduce a factor of safety against tendon rupture and against failure of individual elements of the tendons.

  (b) The following stress allowables will enable the Designer to compensate for elastic and friction losses in the containment during tendon stressing.

      A single tendon shall not exceed a stress of $0.75f_{pu}$ at the anchor. This value should consider ram calibration accuracy as follows: f not greater than f_{pu} (0.75 – ram accuracy). Example: If calibration accuracy is +/– 2%, then the maximum value is $0.73f_{pu}$

  (c) The average tension stress of all tendons after completion of prestressing shall not be considered higher than $0.70f_{pu}$ for design.

  (d) The effective prestress shall consider losses from the following and any other losses applicable to the system as discussed in CC-3542:

    (1) Anchorage slip

    (2) Elastic shortening

    (3) Concrete creep and shrinkage

    (4) Steel relaxation

    (5) Friction losses

CC-3440 Concrete Temperature

If the limits specified in this Subarticle are not exceeded, the material properties will not be significantly altered and no deterioration of the concrete is expected. The Code permits higher limits, provided tests are performed to evaluate the reduction in strength and possible deterioration of the concrete.

Short term is for unexpected events in which the plant would be shutdown and subsequently inspected to verify the structural adequacy of the affected member.

Long term is for normal operating conditions.

The temperature limits in this section basically limit the amount of compressive membrane strain that the liner plate will be subjected to which is about 0.002in/in (mm/mm).

CC-3500 CONTAINMENT DESIGN DETAILS

CC-3510 Design for Flexure, Axial, and Shear Loads

CC-3511.1 Factored Load Design Detailed designs are illustrated in (b), (c), and (d). These sub-paragraphs are similar to the strength design method described in various Editions of ACI 318.

Subparagraph (e) describes the relationship between the concrete compressive stress distribution and the concrete strain which may be assumed in the analysis of sections. ACI 318 allows the assumption of a "rectangular, trapezoidal, parabolic, or any other shape that results in prediction of strength in substantial agreement with results of comprehensive tests." CC-3511.1 however, specifically allows the use of a triangle or parabola; other assumptions would have to be proven by comprehensive tests to be consistent with the prediction of stress and strains. (See Figure CC-3421-1.)

CC-3511.2 describes the straight line theory of stress and strain for service load design. This method is the same as the "Alternate Design Method" described in various Editions of ACI 318 when this Code was originated (previously referred to as the "working stress" method).

CC-3521.1 Tangential Shear

1.0 General The development of the ASME Section III, Division 2 Code tangential shear provisions took several years for implementation. A report was prepared which basically became the basis for what eventually was put into the code. This report is included in Appendix E as [E-1]. The material in [E-2] is a report to ASME-ACI 359 Subgroup on design, TANGENTIAL SHEAR CODE PROVISIONS by Task Group on Shear. The task group report also became the basis for what eventually was put into the code.

CC-3521.1.1 Reinforced Concrete This subparagraph gives design equations to be used in determining the amount of reinforcing steel necessary to resist combined tangential shear and membrane forces. CC-3421.5 must be used in conjunction with this subparagraph when determining the amount and type of reinforcing. When the tangential shear force is not high, then only orthogonal (two-way) reinforcing is required (Equations 12 and 13) with the inclined steel taken as zero. However, when the tangential shear is high, inclined reinforcing is required and the amount of total reinforcing is given by Equations 12 and 13.

The determination of whether the applied tangential shear is high or low is made by applying the provisions of CC-3521.1.1 (b) (1). A design example is shown in Appendix D item [D-2]. For the derivation of Equations 12 and 13, see the two items in Appendix E.

Subparagraph CC-3422.1 permits yielding in the inclined reinforcing, however, the strain is limited to $2\varepsilon_y$.

CC-3521.1.2 Prestressed Concrete The allowable tangential shear stress resisted by the concrete is defined in Article CC-3421.5.2. This Code section says that the minimum prestress shall meet the requirements of CC-3521.1.2, Prestressed Concrete, which requires the following:

(a) A sufficient amount of prestress shall be provided so that N_h and N_m are negative (compression) or zero. Thermal membrane forces shall be included in N_h and N_m for the calculation of effective prestress.

Item(c) further states "When the section under consideration does not meet the requirements of either CC-3521.1.2 (a) or (b), additional reinforcement shall be provided according to requirements of CC-3521.1.1". The requirements of CC-3521.1.1 could lead to having to provide inclined reinforcement. However this is most likely only due to a very large earthquake.

The Code states that orthogonal reinforcement may be used without inclined reinforcement up to a tangential shear stress of $0.2*[f_c']$; however, the US NRC, NUREG-0800 (Reference 11) has placed a limit of $10*[f_c']^{0.5}$. The design should consider several options and determine the most beneficial design.

By both options the following is meant:

Use a larger prestress level so that the concrete can resist some or all the applied tangential shear loads, or use a lower prestress level and additional normal reinforcing to resist the applied tangential shear loads.

There are basically three ways to design a Prestressed Concrete Containment
Vessel [PCCV]:
Provide enough prestressing so that

$$V_c => V_u$$

Provide enough prestress so that

$$V_c + V_s => V_u$$

Use less prestress and let

$$V_s => V_u$$

Cases (a) and (b) rely on concrete in tension but require that liner accident thermal forces be considered. This thermal effect leads to higher prestress, consequently thicker walls, and then higher earthquake forces. With these cases, increased prestress is required in areas such as the dome apex where it is really not needed.

Now case (c) basically uses the approach to prestress for accident pressure loads and reinforce for earthquake loads.

Case (c) may be preferable. It does not rely on concrete in tension for load resistance and it is not subjected to the assumptions that must be made as to just what is the thermal force due to accident conditions when combined with other loads.

Appendix E, Tangential Shear Considerations, has more information on this subject.

CC-3521.2 Radial Shear

CC-3521.2.3 Shear Reinforcement (a) This subparagraph continues the practice of designing "web" reinforcement to carry the excess shear stress over that permitted for the concrete of an unreinforced web. Upon thorough revaluation of research, ACI-ASCE Committee 426 [Reference 2] found that such an approach is desirable.

(b) The method for computing the required amount of web reinforcement is based on the truss analogy. This was recommended by ACI-ASCE Committee 426 after reevaluation of test findings. The derivation of Equations (17), (18) and (19) is given in many textbooks.

(e) This provision regarding the maximum spacing of web reinforcement has also been used in the ACI-318 Code.

CC-3532 Reinforcing Steel Splicing

CC-3532.1 Tension Splices

CC-3532.1.2 Development Length In a location subject to biaxial tension, development length is increased by 25%. This is a departure from ACI-318, where requirements are based on a uniaxial stress situation. The added 25% is a measure of conservatism deemed necessary because of lack of test data of the biaxial stress situation.

CC-3540 Prestressed Concrete

CC-3541 General

Assumptions are provided for use in service load investigations and for review of sections at transfer of pre-stress forces. This section does not apply to the design of compression members in general, but only to members that are prestressed.

In considering the area of the open ducts, the critical areas should include those having coupler sheaths that may be of a larger size than the duct containing the tendon. Also, in some instances, the trumpet or transition piece from the conduit to the anchorage may be of such a size as to create a critical area.

CC-3542 Loss of Prestress

The causes for loss of pre-stress are listed. For an explanation of how to compute these losses, see the reports of ACI-ASCE Committee 423 (Reference 5) and ACI Committee 435 (Reference 6).

Data (Reference 5) have been assembled and analyzed to permit computation of the stress loss due to relaxation of tendons composed of stress-relieved wires. Subsequent work on stress-relieved strand conforming to ASTM A 416 indicates relaxation losses of about the same magnitude.

Stabilized strand or wire is material that has smaller relaxation losses than conventional stress-relieved material. While the strand or wire is at the elevated temperature used for the stress-relieving operation, it is subjected to a high tensile force, which produces a specific amount of permanent elongation, thus resulting in low relaxation losses after the tendon is put into service. For specific relaxation values of a particular steel, the Designer should consult the steel manufacturer. In determining relaxation losses the expected operating temperature at the tendon location in the shell should be used.

Friction losses due to wobble and curvature can be computed by Equations (22) and (23) of the Code, CC-3542. Guidance on the friction that can be expected with particular tendons and particular ducts can be obtained from the manufacturers of the tendons. Over-estimation of the friction may result in extra prestressing force if the estimated friction values are not attained in the field. If the estimated friction factors are determined to be less than those assumed in the design, the stressing force should be adjusted to give only that theoretical prestressing force in the critical portions of the structure required by the design.

CC-3543 Tendon End Anchor Reinforcement

Some bonded reinforcement is needed in the tendon end anchor zone to provide concrete confinement and prevent uncontrolled concrete cracking. The amount of reinforcement required in CC-3543(a) is considered adequate for tendons in the 1000 to 3000 kip (4.48 to 13.45 MN) range; however, this reinforcement requirement may be reduced based on testing.

Appendix F has a report which documents testing that has been performed on simulated full scale tendon end anchor zones for large capacity tendons. The test reports are also in References 9 and 10. Care must be taken in providing reinforcement, since too much may lead to improper concrete consolidation and voids under the bearing plate may result.

CC-3544 Curved Tendons

Curved tendons result in forces applied normal to the tendon and can be determined by the equation:

$Q = F/R$

Q = normal force from the tendon (kips/in.)

F = force in the tendon (kips)

CC-3545 Radial Reinforcement

It is required that radial reinforcement be placed in the dome of prestressed concrete containments, and considered in the cylinder, to maintain the integrity of the structure in the event that delamination planes form parallel to the surface. The spacing of the radial reinforcement should be as required by CC-3545.

The following equation may be used to size the radial reinforcement:

$$A_s = [x_b * Q * A_c]/[0.67 * f_y * t]$$

Where:

A_s = area of radial reinforcement (in.^{2})

A_c = area of concrete for each radial bar, equal to the spacing squared, if a square pattern is used (in.^{2})

t = thickness of the dome or cylinder (in.)

x_b = distance from the outside face to the centroid of the prestressing tendon group (in.)

Q = the normal pressure after transfer of prestress due to all of the dome or cylinder tendons (psi)

f_y = yield strength of radial reinforcing

DELAMINATION HISTORY AND PREVENTION

After the delamination of a Nuclear Power Plant prestressed concrete dome, provisions were incorporated in the ASME Section III, Division 2 Code to:

(1) Provide criteria that would have a high probability of preventing delaminations from occurring using sufficient concrete compression allowables.

(2) Point out the need to use net concrete section areas for unbounded tendons in computing compression stresses.

(3) Require radial reinforcement that would maintain the entire concrete thickness to function as a unit, in the unlikely event that the concrete shell would form delamination planes.

TENDON

F

Q

R

-F

Sufficient concrete must be provided between
the tendons and the inside of the containment
and at penetrations to prevent a punch-out
failure as shown below.

Q

Q

FAILURE
PLANES

Tendon Displaced Around Opening

Centerline of Opening

HOOP TENDON
RADIAL FORCES

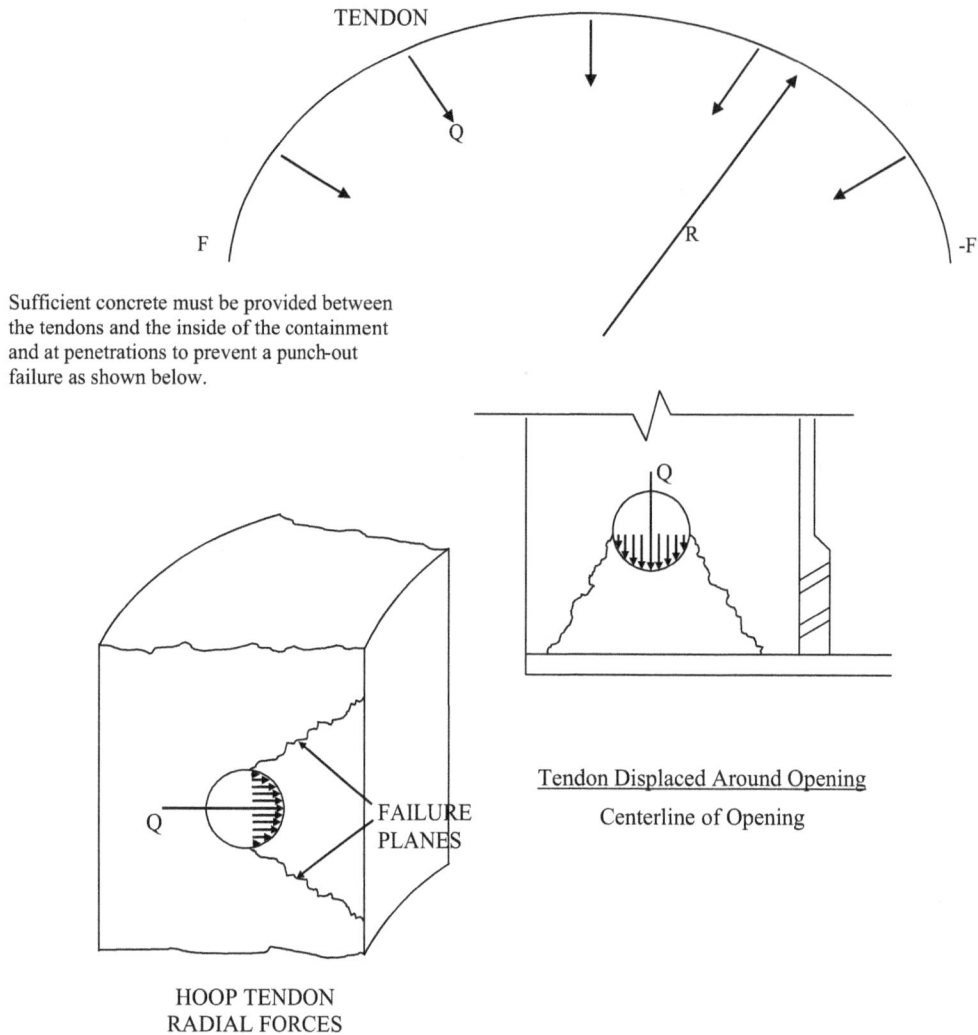

FIGURE CC-3545-1   RADIAL REINFORCEMENT.

Historical Dome Delaminations

(A) The first Containment Structure in the United States to experience a dome delamination was the Turkey Point Unit 3 Containment in 1970.

Near completion of prestressing on the Containment Dome, it was noticed that there was some outer surface concrete spalling at a radial location of about 30 to 40 ft from the apex as concrete was removed to determine the extent of the spalled or what became known as the delaminated concrete (cracking/separation essentially parallel to the surface). At the apex the depth was about 12 in from the outer surface. The dome thickness was 39 in. An extensive investigation was performed to determine the cause and what should be done to restore the dome to an acceptable condition. The design concrete strength based on 6" by 12" cylinders was 5000psi.

INITIAL CAUSE INVESTIGATION

The initial preliminary investigation revealed the following:

Several construction joints used to layers of expanded metal which created some voids.

Construction did not follow the Engineers' specified prestressing sequence and this resulted in some degree of unbalanced loading at various stages.

FINAL CAUSE INVESTIGATION

An extensive investigation was done to determine the true strength of this large shell with inplane biaxial compression combined with radial tension in the outer portions of the dome thickness with sustained loads and unbonded tendons.

Some of the items examined were:

Due to the fact that 5 layers of tendons (with 4" diameter ducts) were used, what was the effective reduction of cross section, since these were not grouted and used a corrosion inhibiting filler? This reduction was determined to be 25 percent based on testing models of the actual removed concrete.

Based on reviewing test results of large cylinders (24" by 48") compared to normal (6" by 12") it was concluded that a large concrete member would have a strength of only about 80 percent of a typical concrete cylinder.

The tests performed on both normal and large specimens had end restraint in the form of friction which leads to increased strength and an apparent 45 degree shear type failure surface.

To investigate the effect of end restraint, which would not exist in long members or shells, tests were performed on cylinders with Teflon wedge-shaped layers on the top and bottom to minimize friction. These tests showed another strength reduction of about 20 percent. Also the 45 degree failure surface did not exist. The cylinders split in the vertical direction apparently due to a Poissons' strain failure effect. This was interesting since this is what appeared to happen in the actual structure being investigated.

Another question had to do with sustained loading compared to the short term normal cylinder tests. Specimens were tested and found to fail at about 90 percent of the short term normal test values.

Direct tension tests were done on both delaminated concrete and 6" by 12" cylinders from the jobsite storage. These were actual tension tests and not splitting tensile tests. These tests indicated that the concrete tensile strength at this plant was about 6 percent of the 6" by 12" compressive cylinder tests. Concrete from other jobsites indicate that the tensile strength was typically 10 percent of this value. Six percent of 5000psi is 300psi, however, if we apply the same logic relative to strength reduction from the missing area due to ducts and other effects to the tension strength, it is most likely about 180psi.

Combining all the above items that affected compressive strength only it was concluded that a large compression member that did not have end restraint with sustained loading and 25 percent of the area missing would have a compressive strength of,

$$(0.75)*(0.8)*(0.8)*(0.9) = 0.432.$$

Hence a 5000psi concrete that is now, effectively

$$(0.432)*5000 = 2160psi.$$

If one uses a straight line interaction between compression and tension and considers that the applied initial prestress would be about 100psi then the tension around 12" below the outer surface will be about $[12/39]*100 = 32psi$. The compression that would be associated with this value would be $[(180-32)/180]*2160 = 1776psi$.

The calculated membrane stresses in the outer portion of the dome were about 1750psi, and with some temperature gradient, even higher. Therefore, it was concluded that all the previous items added up to a condition where the dome was stressed to its ultimate capacity in the outer portion. About 12" in the concrete was in triaxial compression and had a very high strength.

As shown above, the applied loads basically equaled the structural capacity; however there was some contribution of the concentrated tendon radial force.

In summary, the items listed above were the main contributors to the delaminated condition. The construction joints were also a contributor.

(B) The second Containment Structure in the United States to experience a dome delamination was Crystal River Unit 3 [CR3] in 1976. In 1976, approximately 2 yrs after concrete placement and 1 year after tensioning, the 3 ft thick containment dome at CR3 delaminated. Potential contributing factors were investigated in an attempt to determine the cause or causes of the delaminated condition. Several effects which may have contributed to

the problem were identified. It was concluded that radial tension stresses combined with biaxial compression stresses initiated laminar cracking in concrete having lower than normal direct tensile capacity and limited crack-arresting capability (friable particles). Reference 14 provides details of the delamination, root cause, and repair strategy.

(C) Containment in India. At this Containment in 1994, the inner containment delaminated and since it had no liner plate, pieces of concrete actually fell inside the containment. This failure occurred when only about 36 percent of the tendons were stressed. This certainly indicates a major amount of under-design. This dome was rather thin and even had large openings for large equipment removal. Therefore, this led to increased loads in the dome because of tendon layout. For more information refer to Reference 15.

(D) The first Containment Structure with delaminations in the cylinder.

A more recent case of delamination occurred in 2009 at Unit 3 of Crystal River Prestressed Containment located in Florida. As part of the Steam Generator Replacement project, a cut was made in the containment wall above the equipment hatch for removal and replacement of the existing Steam Generators. The cut was made using the hydro-demolition method, which utilizes high pressure water jets. Prior to the removal of concrete, the tendons in the construction opening, vertical and horizontal tendons were detensioned and removed. During the removal of the concrete at the construction opening, gaps from delamination were observed near the hoop tendons. This gap was not anticipated and, based on industry operating experience, other similar projects have not encountered the same condition. Extensive cracks were observed in the construction opening area. The main cracks were observed at the construction opening area in the plane between the horizontal and vertical tendons and parallel to the surface of the wall. The cracks seemed to go right through the aggregates, indicating the cement paste may have been stronger than the aggregates.

The root cause as of November 17, 2009 determined the following were the combined contributors: (a) Inadequate Containment Cutting, (b) Inadequate concrete–tendon Interactions, (c) Shrinkage, Creep, and Settlement, (d) Chemically or Environmentally Induced Aging, (e) Inadequate Use of Concrete Materials, (f) Inadequate Concrete Construction and (g) Inadequate Concrete Design due to High Local Stress.

Reference 4 provides details of the delamination, root cause, and repair strategy.

CC-3600 LINER DESIGN ANALYSIS PROCEDURES

CC-3610 General

Here is a statement that says components should be evaluated for cyclic loading. This is difficult to do on the liner anchors without using tests.

Appendix G has the following papers: [G-1] Liner system design example, [G-2] Containment Liner Plate Anchors and Steel Embedments Test Results and also includes cyclic effects on anchors and also shows results for various types of anchors, [G-3] Increase in steel liner strains due to concrete cracking, a reexamination of the tension liner allowables and [G4] Liner Anchors Interaction of Tensile and Shear Forces.

CC-3620 Liner

The liner (and anchorage) analysis must consider the worst case deviations in liner geometry due to fabrication and erection tolerances as specified in the Code and the construction specification.

The tolerances stated in CC-4522 and Nonmandatory Appendix D of the code is mainly applicable to the general shell shape. Table CC-4523-1 does state the requirements for maximum joint offset, which, in the case of a 1/4" thick liner plate, will be 1/16". The NUREG/CR-6810 (Reference 13), which documents the Sandia testing of 1/4 scale model liner plate samples, indicated basically no reduction in strength due to an offset of this magnitude. However, the total elongation was reduced by about 50%. These tests were for tension. Compression would most likely have less of an effect. Now, since liner plates have ultimate strain values of about 30%, joint offset reduction should not have an effect on the strength under specified design conditions.

One of the most important deviations from the ideal liner geometry is the existence of inward curvature between liner anchors. This value should be defined in the fabrication and erection/construction project specification. Deviations, greater than those considered in the design, found during construction should be justified by the designer or repaired.

CC-3630 Liner Anchors

An ongoing problem with liner plate material is that it may far exceed the ASTM minimum specified yield. A project could specify a maximum yield but that could end up becoming very expensive. A more practical solution is to use higher capacity, ductile anchors. The ductility really comes from having enough anchor steel shear strength to be able to locally crush the concrete. This crushing is fully acceptable since it will only happen once, in the unlikely event of a nuclear accident.

CC-3700 LINER DESIGN

CC-3710 General

Since this section shows the acceptance of using test results to establish acceptable allowables it is probably acceptable to even provide test results to justify liner anchor spacing that exceeds the calculated values in Table CC-3720-1 under construction loads. However the test results will be required.

CC-3720 Liner

Construction allowables have been included because temporary loads during construction can be more detrimental to the liner than service loads. These loads should not be overlooked. The setting of allowables for this condition is intended to call it to the attention of Designers and Constructors to prevent damage to containments during construction.

The principal consideration in setting strain allowables was that the plate steels allowed for construction of liners have a minimum elongation of 20% in an 8-in. gage length. Steels in this category can tolerate large strains or displacements without violating the leak tightness of the structure.

This section basically says that basically the loads imposed by the concrete placement on the liner plate really only impose a rather small strain since, as the concrete sets up, there is no potential for any further inward displacement and strain and/or stress in the liner plate. Construction-related deformations are, as an example, those imposed on the liner when bringing sections into alignment and other normal erection activities. Severe erection activities would include things such as: excessive grinding causing thinning, poor welding practices, excessive arc strikes, etc.

CODE TABLE CC-3720-1

(A) "The stress allowables of 2/3 f_{py} were set in accordance with normal structural allowables and were for construction type loads such as concrete form pressures. Using test results to establish acceptable allowables is probably acceptable to justify liner anchor spacing that exceeds the calculated values in Table CC-3720-1 under construction loads. However, the test results and a Code Case or change will be required."

(B) Service Category strain allowables were set to be conservative as stated in the Code and they are defined as:

"NOTES: The types of strains limited by this Table are strains induced by other than construction-related liner deformations."

These strains are self limited under the code design loads since the liner is attached to the concrete.

(C) Factored Category strain allowables were also set to be conservative, as stated in the Code, and they are defined as, with units of in/in [mm/mm]:

Membrane Compression =	0.005
Membrane Tension =	0.003
Combined Membrane and Bending Compression =	0.014
Combined Membrane and Bending Tension =	0.010

The plate material used for liner plates may have a specified minimum yield as low as 30 ksi and a yield strain of 0.001 in/in (mm/mm). However, past experience has found that the actual yield values may be significantly higher.

The mean ultimate uniaxial tensile strain is about 0.30 in/in (mm/mm) with a biaxial reduction factor of 0.50, this results in a strain capacity of 0.15 in/in (mm/mm). This is about 10 times higher than the combined membrane and bending compression allowable stated above and 15 times greater than the combined membrane and bending tension value stated above. The tension value was set at a more conservative value than the compression value since liner failure would most likely result from a tension condition.

For membrane only, the strain of 0.15 in/in (mm/mm) is about 30 times higher than the membrane compression allowable stated above and 50 times greater than the membrane tension value stated above. The tension value was set at a more conservative value than the compression value since liner failure would most likely result from a tension condition.

The allowables are to be considered conservative when compared to the liner capability. As illustrated in Appendix G, the computed compressive strains will probably not exceed 0.002 in/in (mm/mm).

The allowables are considered conservative when compared to the liner capability. In the example given in Appendix G, the computed compressive strains do not exceed 0.002 in/in. (mm/mm)

CC-3730 Liner Anchors

Liner anchor allowables are divided into two categories. Mechanical loads are those in which movement or adjustment of the structure during testing or operation will not relax the condition causing the load. Examples of loads of this type are anchors where equipment is attached.

The second load category is displacement-limited loads. Most anchor loads probably fall into this category. Examples of loads in this category are loads imposed by temperature changes or displacement changes in the concrete structure during the life of the plant. Allowables in this category are given in ultimate capacities. Normally, the ultimate capacity of the anchor will be determined by testing. Figure CC-3730-1 shows the definition of the ultimate capacity of the anchor.

In Table CC-3730-1 Liner Anchor Allowable, Force and Displacement Allowable are stated. Force allowables are for mechanical loads and these were set based on structural steel allowables being used when the code was written. The displacement values were based on the experience gained doing design and analysis as illustrated in the Appendix G design example. The design example used an energy design concept for acceptability. This was converted to an allowable displacement technique by the containment designer for easier understanding.

There is some information in the Sandia NUREG/CR-6810 (Reference 13). This document shows that the primary effect of weld offset of 25 percent of the thickness is an approximate 50 percent reduction in ultimate elongation when considering tension.

CC-3740 Penetration Assemblies

This section references ASME Section III, Division 1. However, Division 1 does not provide design information on the anchorage of a penetration assembly into a concrete containment. CC-3000 must be used for this aspect of the design. When considering impulse and/or impact loadings, there are increased allowables and displacements that may exceed yield up to the ductility limits defined in CC-3923.

CC-3750 Brackets and Attachments

This section references the AISC Manual for Steel Construction. However, AISC does not provide design information on the anchorage of a Bracket or Attachment into a concrete containment. CC-3000 must be used for this aspect of the design. When considering impulse and/or impact loadings there are increased allowables and displacements that may exceed yield up to the ductility limits defined in CC-3923.

ACI 349-06 and ACI 349-07 have some good information on Embedment design and are useful as supplemental material. However, the requirements of ASME Sec III, Division 2 must always be satisfied.

Examples

The term "embedment" covers a broad scope which includes anchors, embedment plates, shear lugs, and specialty inserts.

The anchors' strength design shall satisfy the following, where applicable:

- Steel strength in tension
- Steel strength in shear
- Concrete breakout strength of the anchor in tension
- Concrete breakout strength of the anchor in shear
- Pullout strength of the anchor in tension
- Concrete side-face blowout strength in tension
- Concrete pry-out strength of the anchor in shear
- In all the above, the combination of shear and tension anchorage design

CC-3800 LINER DESIGN DETAILS

CC-3810 LINER ANCHORS

In most cases, when the actual yield strength of the material used in a structure exceeds the minimum value of the ASME material specifications, the result is a conservative design. However, in a liner subjected to a temperature increase, the maximum temperature expansion load from the liner is increased for liner yield strengths greater than ASME guaranteed minimum values. Therefore, the conservative approach is to consider that the liner remains elastic and the forces are not limited by yielding.

Many of the design and analysis effects that should be considered are illustrated in the design example in Appendix G.

There is also some information in the Sandia NUREG/CR-6810 (Reference 13). This document shows that the primary effect of weld offset of 25 percent of the thickness is to get about a 50 percent reduction in ultimate elongation when considering tension.

Concrete voids behind the liner should not be a problem. Tests were performed circa 1975 that subjected a 1/4" plate with a span of over 24" to a pressure of 60 psi. In the final loaded condition the plate was a catenary. Unfortunately, these tests cannot be located at this time. Also, with all the containments that have been tested, it appears that none have reported voids behind the liner as a problem.

During the Sandia Reinforced Containment 1/6 Scale Tests the studs actually failed the liner plate, resulting in tears. There was apparently a problem finding small studs to properly scale the full size and larger ones were used relative to the scaled liner plate thickness. Therefore in using stud anchored liner plates, tests should be performed to prove that the stud will shear off and not tear a hole in the liner plate.

As an alternative to the above technique, biaxial compression tests may be performed and used to define the maximum limit of yield strengths.

CC-3830 Transitions From Concrete To Steel

These paragraphs provide guidance in those areas where the structure of the pressure boundary changes from concrete to steel. Examples of this case are where the steel pipe of a nozzle passes through the wall of the containment, the installation of an equipment hatch, or large access openings, or the case of a full diameter transition where the head or upper portion of containment is steel for economic or access reasons.

The distance of 25t, for which the metal shell thickness shall not be reduced, at junctions of metal to concrete containments, is considered an adequate minimum in current practice.

CC-3900 DESIGN CRITERIA FOR IMPULSE LOADINGS AND MISSILE IMPACT

There is some good information on Impactive and Impulsive loads in ACI 349-06, Appendix F. However all the provisions in ASME Section III, Division 2 must be satisfied.

IMPACTIVE: Impactive loads are time dependent loads due to collision of masses. Impactive loading may be defined in terms of time dependent force or pressure. Examples of impactive loadings include tornado generated missiles, whipping pipes, aircraft missiles, and other internal and external missiles.

IMPULSIVE: Not associated with collision but includes jet impingement, blast pressure, compartment pressurization, and pipe whip restraint reactions.

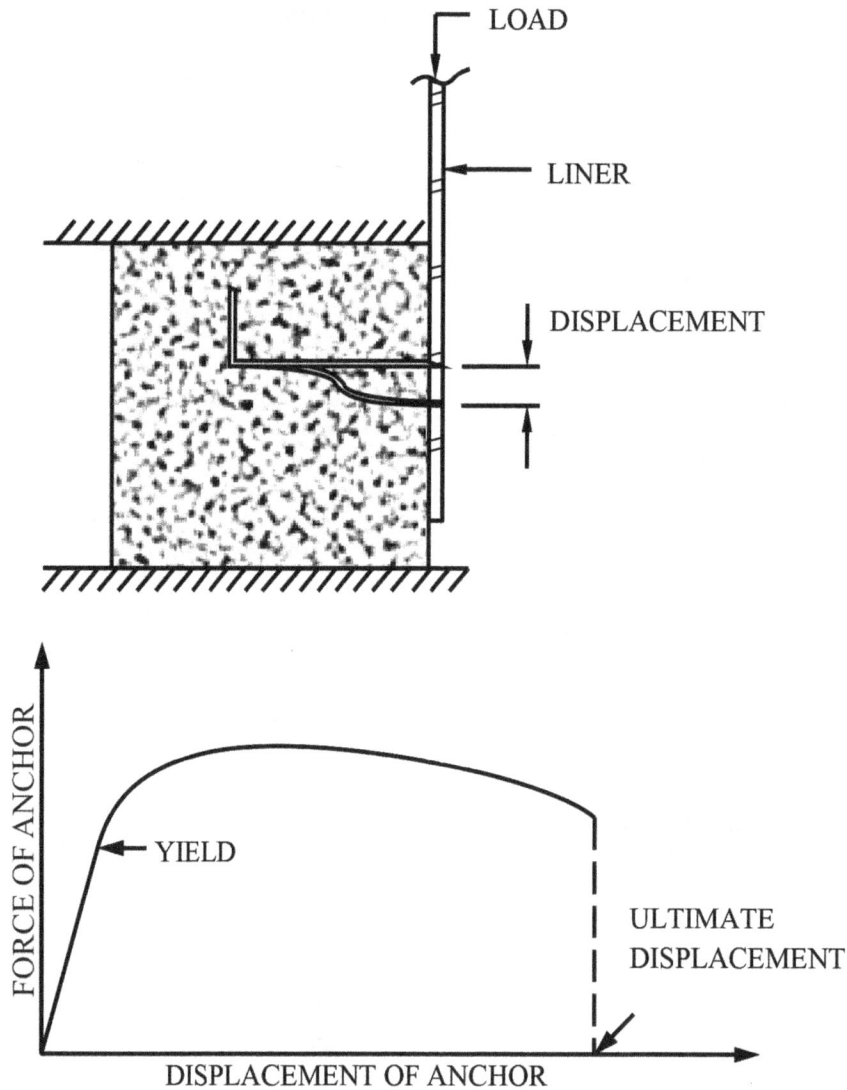

FIGURE CC-3730-1 ULTIMATE CAPACITY OF THE LINER ANCHOR.

The Code recognizes the general requirement to assure containment integrity in the event of an impulse or impact load resulting from Abnormal or Extreme Environmental effects. It further recognizes that such loadings are dynamic in nature and their effects tend to be localized. The Code also recognizes the nature and magnitude of such loads requires that relatively large amounts of energy must be absorbed by the containment structure. To permit this energy absorption in a rational design, implicit plastic response of the structure is permitted. The amount of plasticity permitted is controlled by the ductility factor, μ, as limited in CC-3923.

However, it is noted that the valve actuation impulse load category (G) may be in the normal load category. In such an instance, the above discussion does not apply and service category allowables are to be used for (G) in service load combinations.

CC-3922 Stress Allowables

The implicit plastic response of the structure defined by the use of ductility limits in excess of unity has the effect of defining equivalent static load seen by the structure. When the ductility

limits of CC-3923 are used to define equivalent static loads, the applicable code stress limits of CC-3400 or CC-3922 are used to verify design.

At present the Code does not define stress strain limits if more rigorous nonlinear "time history analysis is used.

Both concrete and steel properties are affected by strain rate effects which have been used to increase the limiting f_y and f_c values given in CC-3400. Increases in these design limits, Dynamic increase Factors (DIF), must be verified by test results or authoritative reference as detailed in the Design Report.

CC-3923 Ductility Limits

The ductility limits defined consider that impactive type loads have a limited energy to impart to the containment structure. Since there is an upper bound in the input energy to be absorbed by the structure, this design safety margin for ductility is limited to 1.5. For impulsive loads, slight increase in calculated load magnitudes can have a significant increase in ductility requirements. For this reason safety margin on required ductility is limited to 3.0.

CC-3931 Penetration Formulas and Impulse or Impact Effects

It has long been recognized that in the immediate area of a missile impact failure of concrete typically results and any stresses determined in this area would exceed code allowables. It is necessary to provide limits on such localized areas for design purposes. A damage diameter of ten times the mean diameter of the impacting missile is based on the ratioing of observed yield line pattern diameters to missile diameters for a variety of missiles impacting concrete (Reference 7). The factor of has been defined as the impact area for highway bridges designed to resist impact loads (Reference 8). The value of "t" is expressed in feet.

CC-3932 Effective Mass During Impact

It is well known that transfer of energy to a target structure by an impacting mass is a function of the mass of the target structure which experiences the impact during the time energy is being transferred. The effective mass of the structure is also a consideration when determining the frequency of structural response to the impact. There are, at the present time, considerable differences of opinion on how that participating mass should be defined. The definition in the Code conservatively assumes the participating mass consists only of that portion of the target structure that would act if a punching shear failure of the structure were to occur a distance d/2 from the edge of the loaded area. Other definitions might include the area of the target traversed by a shear or compression wave during the time the missile is penetrating the target or while a stress wave is traversing the missile length and returning.

APPENDIX A—GENERAL CONTAINMENT DEVELOPMENT AND CONFIGURATIONS

[A-1] ADVANCES IN THE ANALYSIS AND DESIGN OF CONCRETE STRUCTURES, METAL CONTAINMENTS AND LINER PLATE FOR EXTREME LOADS. J.D. STEVENSON ET AL, SMIRT 10 CONFERENCE AS A PRINCIPAL DIVISION LECTURE.

SUMMARY

The material presented herein summarizes the progress that has been made in the analysis, design, and testing of concrete structures. Many of the most outstanding papers presented from SMiRT 1 to SMiRT 9 are referenced and at times conclusions are summarized.

The material is summarized in the following documents:

A] Containment Design Criteria and Loading Combinations - J. D. Stevenson - Stevenson and Associates, Cleveland, Ohio, USA

B] Reinforced and Prestressed Concrete Behavior - J. Eibl and M. Curbach - Karlsruhe University, Karlsruhe, F.R.G.

C] Concrete Containment Analysis, Design and Related Testing - T. E. Johnson and M. A. Daye - Bechtel Power Corporation, Gaithersburg, Maryland, USA

D] Impact and Impulse Loading and Response Prediction - J. D. Riera - School of Engineering - UFRGS, Porto Alegre, RS, Brazil

E] Metal Containments and Liner Plate Systems - N. J. Krutzik - Seimens AG, Offenbach Am Main, F.R.G.

F] Prestressed Reactor Vessel Design, Testing and Analysis - J. Nemet - Austrian Research Center, Seibersdorf, Austria and K. T. S. Iyengar - Indian Institute of Science, Bangalore, India

[A-2] HISTORY OF THE DEVELOPMENT OF UNITED STATES PRESTRESSED CONCRETE CONTAINMENTS (PCC) AS DESIGNED BY THE BECHTEL CORPORATION, THEODORE E. JOHNSON, JUNE 17, 2008.

HISTORY OF THE DEVELOPMENT OF UNITED STATES PRESTRESSED CONCRETE CONTAINMENTS [PCC] AS DESIGNED BY THE BECHTEL CORPORATION

Theodore E. Johnson
TJBG Consulting Inc.
June 17, 2008

OVERVIEW

This document will review what the PCC development was in the United States up to the termination of new Nuclear Plants about 25 years ago. This document only covers Prestressed Concrete Containments (PCC) that were designed by Bechtel Corporation. Bechtel designed most of the PCCs and led the country in design innovation and development of new design techniques. In addition some information will be stated as to what new items should be considered for future plants for a new generation of PCCs.

This document will only briefly mention containment liner plate since the main topic is the prestressed concrete.

TABLE OF CONTENTS

1   GENERAL OVERVIEW OF UNITED STATES PCC HISTORY

In 1965, the first generation of fully prestressed containments were under design in the United States. They used a cylinder and a shallow dome with a ring girder. The effective prestress was set at 1.5 times design accident pressure [Pa] and the tendons had a capacity of about 500 tons. The horizontal tendons were anchored at 6 vertical buttresses. Nuclear Plants such as Palisades and Turkey Point used this type of design. See Appendix A for some Containment Figures which illustrate overall configuration and tendon layout.

The next generation of containments used an effective prestress level of 1.5Pa or 1.2Pa with 3 buttresses and about 1000 ton capacity tendons. They also used a cylinder and a shallow dome with a ring girder. The Arkansas Nuclear One plant is typical of this generation.

The last generation of containments that were built in the UNITED STATES used an effective prestress level of about 1.2Pa with 3 buttresses and about 1000 ton capacity tendons and the shallow dome was replaced with a hemisphere. The vertical tendons started in the tendon gallery below the base slab. These tendons ran upward over the dome and anchored at the tendon gallery on the opposite side. There was an initial concern as to installing tendons this long, so provisions were made to both pull and stuff the tendons. However, the pulling was sufficient to install the tendons. The Trojan Containment [now decommissioned], was the first to use these "U" tendons.

The tendons used in the UNITED STATES were primarily the BBRV wire tendons. Three containments used a VSL strand system. They were Rancho Seco [now decommissioned], Vogtle and San Onofre.

If the United States Nuclear Power Plant program would have continued, the next generation of containments would have been similar to those described by the papers in the APPENDIX B to this part and discussed below. The one paper describes a Containment that was apparently built in China and designed by an Engineering Company in Finland.

An extensive amount of information is contained in the following documents on the design and analysis of a Prestressed Concrete Containment Vessel and a Steel Liner Plate:

A   BC-TOP-5A PRESTRESSED CONCRETE NUCLEAR REACTOR CONTAINMENT STRUCTURES
B   BC-TOP-1 CONTAINMENT BUILDING LINER PLATE DESIGN REPORT

2   POSSIBLE DESIGN DIFFERENCES FROM THE PAST TO THE NEXT UNITED STATES CONTAINMENT GENERATION

A   A two buttress versus the three buttress design could cause the NRC to classify this design as a prototype and require

extensive instrumentation during the Initial Structural Integrity Test [ISIT].
B   Larger tendons may require justification and may require some testing to provide design proof of adequacy. The latest size available is about 50 percent greater than what has been used in the UNITED STATES. Section 3.3 will address this in more detail.

3   PCC PRESTRESSING SYSTEM

It has been common to go to for bid and allow either wire or strand systems to be proposed by the bidder. Then the low bidder that satisfies all requirements is awarded the contract.

3.1   Tendon Type and Size and Miscellaneous
There have been many tests run on systems in the 1000 ton range, and both wire and strand systems were qualified in the past. Information should be obtained on what tests need to be done on the larger potential systems to satisfy ASME Section Division 2 Code [Code] requirements. Also see comments from Section 2 B]. There have been concerns that relaxation tests have been performed at 70 degrees F and should have been done at a higher value such as 90 degrees F.

3.2   Concrete Strength
Concrete strengths usually were from 5000psi to 6000psi. It is important to get tests run early in the process so that the appropriate concrete information is available for strength, creep and shrinkage.

3.3   Buttress and Concrete End Anchor Design
See Section 2 A on prototype concerns.

3.4   Miscellaneous Considerations
There have been problems with prestressing systems in containments in the past. As an example, there has been wire corrosion due to moisture, anchor head cracking, voids under bearing plates, etc. Steps will need to be taken to avoid these problems in the future.

4   STEEL LINER PLATE SYSTEM

4.1   General Liner Design
The original liner plates consisted of 1/4" plate with 1/4" x 3" x 2" angles at 15" centers running in the vertical direction. The horizontal direction used horizontal channels at about 5 ft centers. A variety of tests were performed to determine the anchor load versus displacement characteristics. Also tests were performed to determine what was called "Bent plate stiffness".

4.2   Anchor Stiffness
It was decided to use structural "Tee Shapes" for the liner anchors for later generations of containments.

4.3   Brackets
A polar crane brackets used normal structural steel. However any major tension members should go completely through the thickened liner plate. This will avoid loading the thickened plate in the through thickness direction. This direction is of lower strength, and there may also be delaminations in the plate.

4.4 Metal Hatch and Locks

The locks and equipment hatch are usually designed by the supplier of these items to the sections of the applicable ASME codes.

4.5 Mainsteam and Feedwater Penetrations

The mainsteam and feedwater penetrations nozzles should penetrate the steel reinforcing plate to avoid through thickness loading. Also since these are high energy pipe lines it is good practice to have the anchorage system basically engulf the entire wall thickness.

CONTAINMENTS

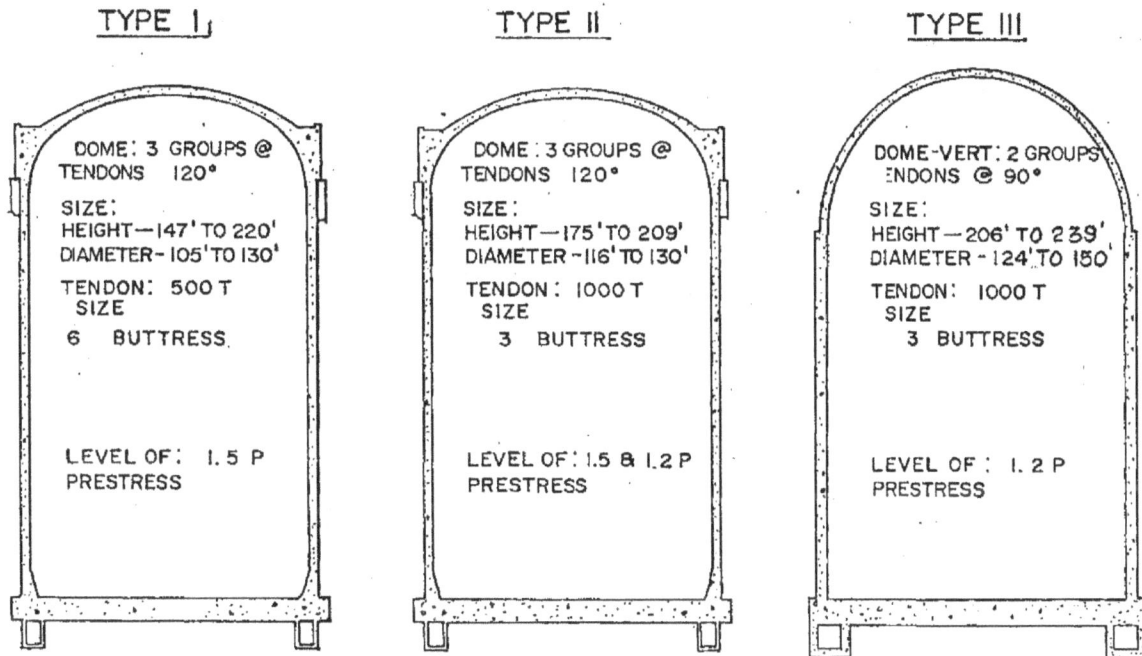

TYPE I

DOME: 3 GROUPS @
TENDONS 120°

SIZE:
HEIGHT—147' TO 220'
DIAMETER- 105' TO 130'

TENDON: 500 T
SIZE

6 BUTTRESS

LEVEL OF: 1.5 P
PRESTRESS

TYPE II

DOME: 3 GROUPS @
TENDONS 120°

SIZE:
HEIGHT—175' TO 209'
DIAMETER -116' TO 130'

TENDON: 1000 T
SIZE

3 BUTTRESS

LEVEL OF: 1.5 & 1.2 P
PRESTRESS

TYPE III

DOME-VERT: 2 GROUPS
TENDONS @ 90°

SIZE:
HEIGHT—206' TO 239'
DIAMETER - 124' TO 150'

TENDON: 1000 T
SIZE

3 BUTTRESS

LEVEL OF: 1.2 P
PRESTRESS

CONTAINMENT HOOP TENDON CONFIGURATION

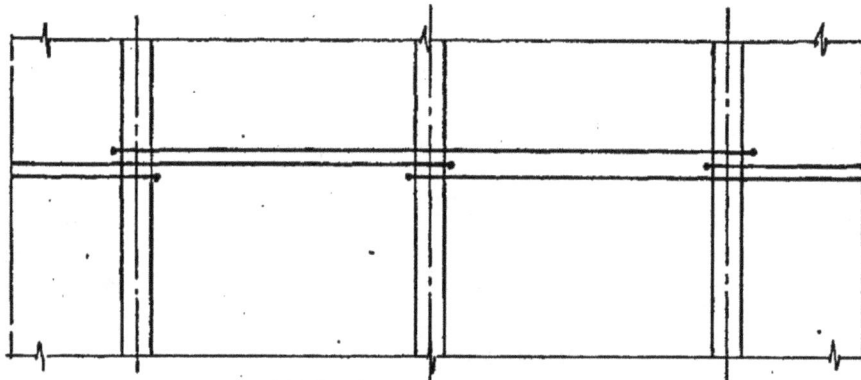

DEVELOPED ELEVATION
HORIZONTAL WALL TENDONS
(3 BUTTRESSES)

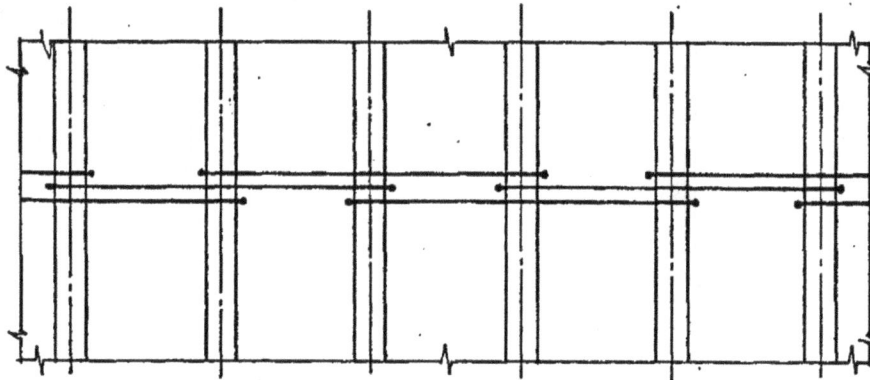

DEVELOPED ELEVATION
HORIZONTAL WALL TENDONS
(6 BUTTRESSES)

CONTAINMENT DOME TENDON CONFIGURATION

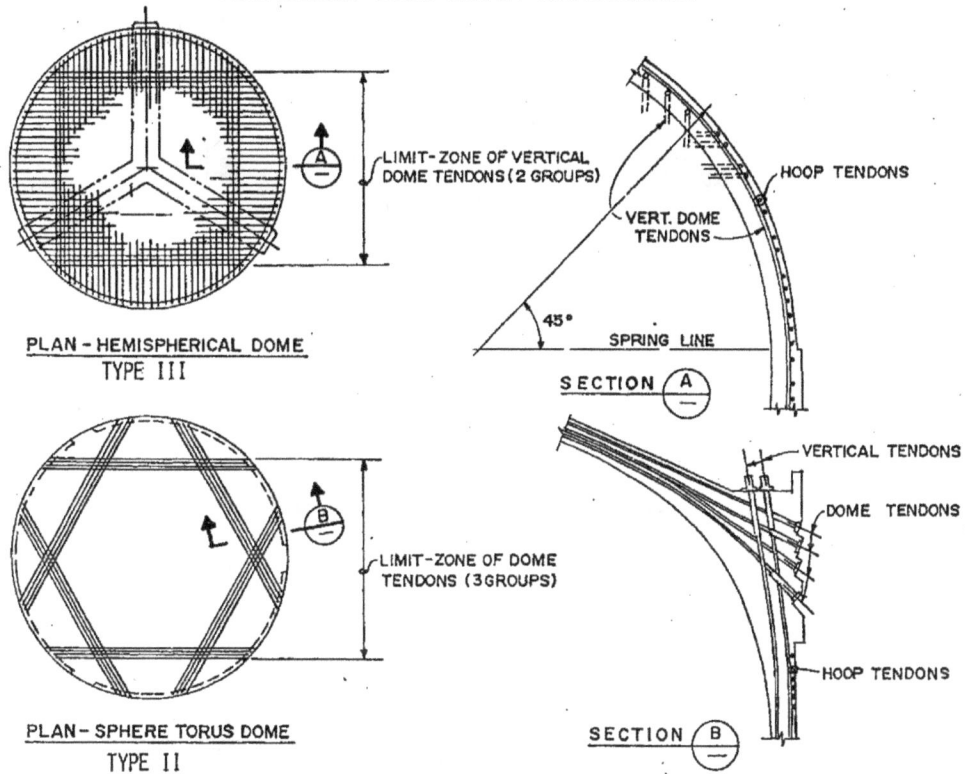

LIMIT-ZONE OF VERTICAL DOME TENDONS (2 GROUPS)

HOOP TENDONS

VERT. DOME TENDONS

45°

SPRING LINE

SECTION A

PLAN-HEMISPHERICAL DOME
TYPE III

VERTICAL TENDONS

DOME TENDONS

LIMIT-ZONE OF DOME TENDONS (3 GROUPS)

HOOP TENDONS

PLAN-SPHERE TORUS DOME
TYPE II

SECTION B

[A-3] FIGURES IN THIS APPENDIX DOCUMENT ILLUSTRATE VARIOUS CONFIGURATIONS OF CONCRETE CONTAINMENTS. THESE INCLUDE BOTH PAST AND THOSE PRESENTLY BEING LICENSED CONTAINMENTS IN THE UNITED STATES.

PAST CONTAINMENT DESIGNS

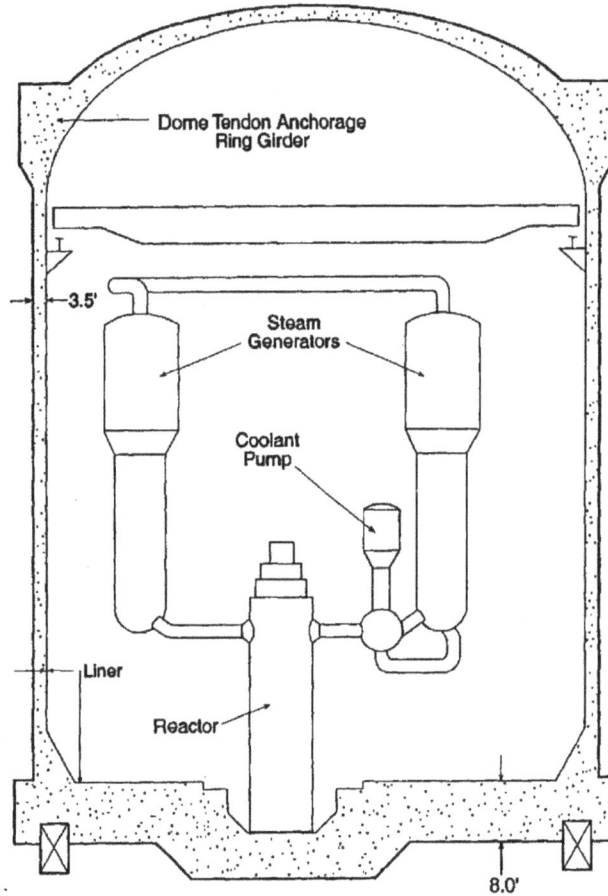

Small PWR Prestressed Concrete Containment

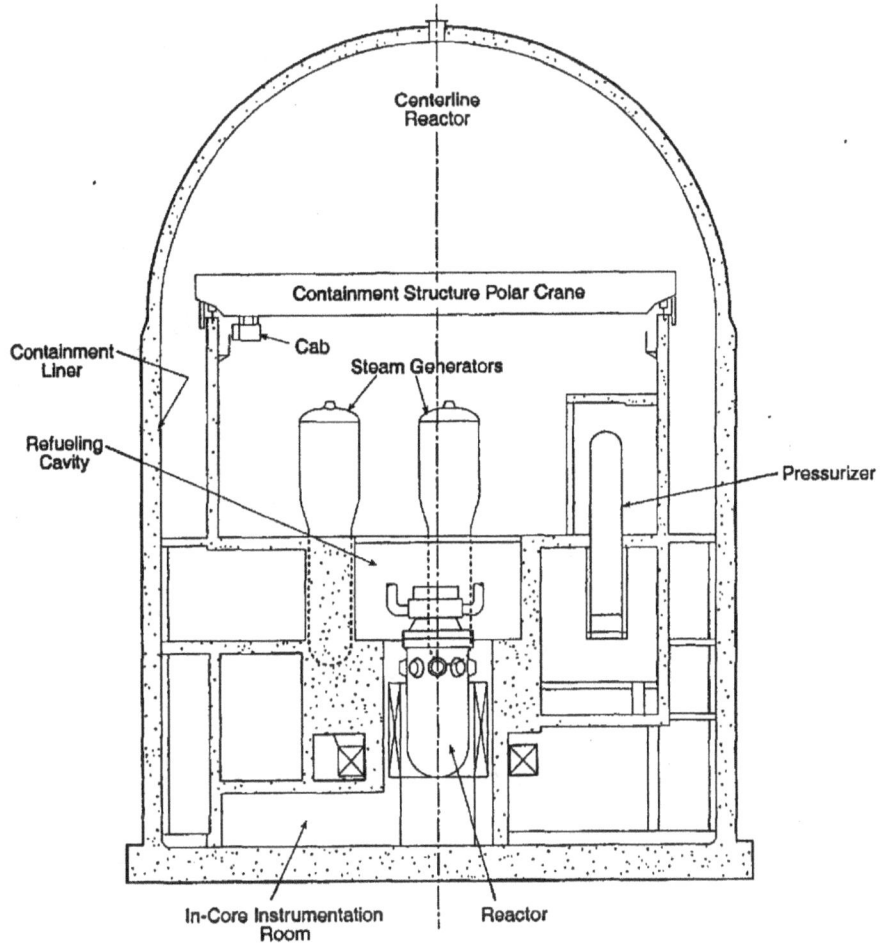

Large or Small PWR Deformed Bar Concrete Containment

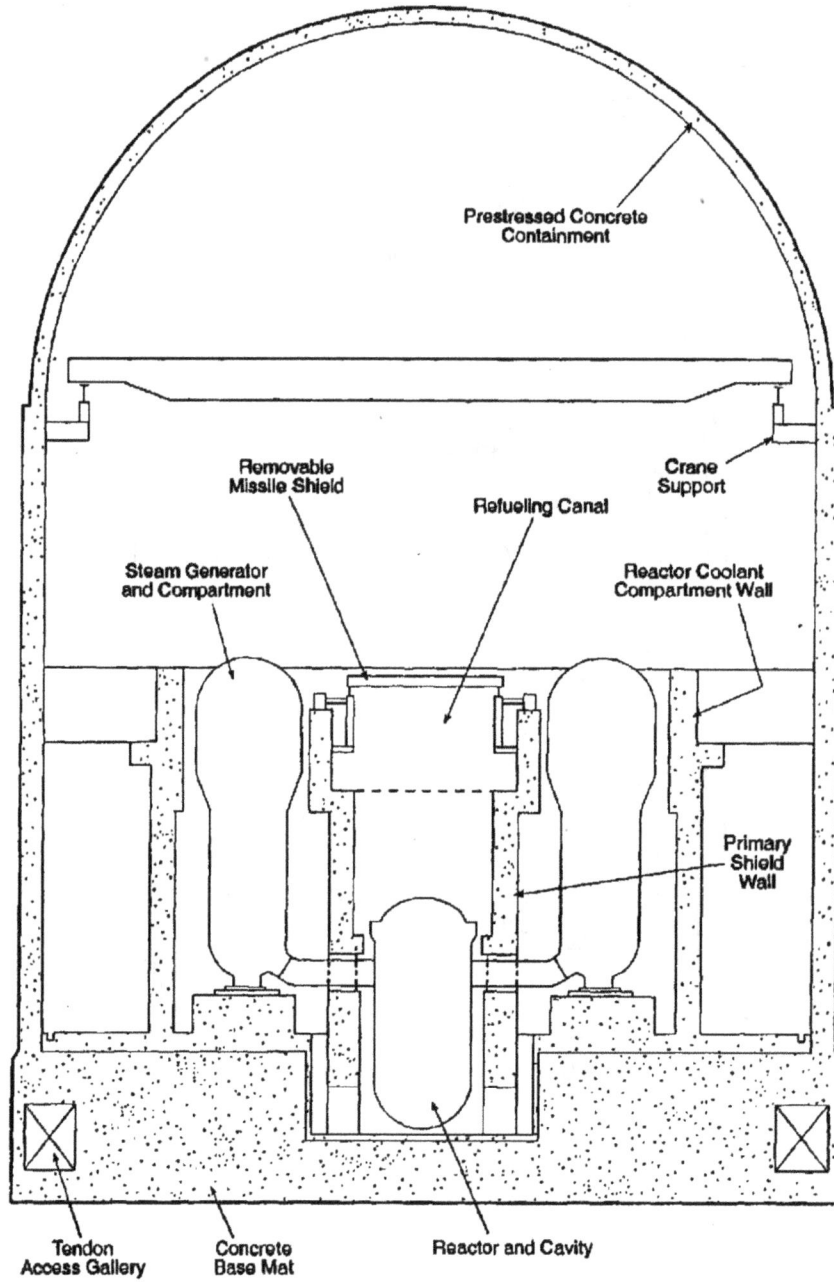

Large PWR Prestressed Reinforced Concrete Containment

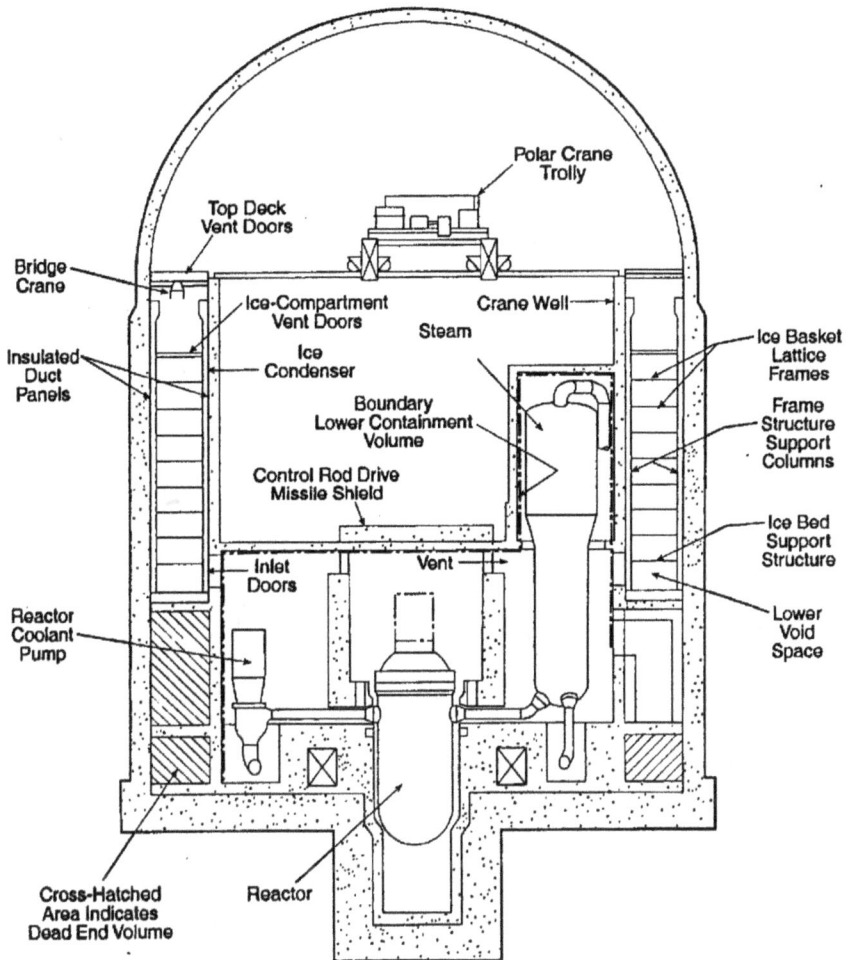

PWR Concrete Ice Containment

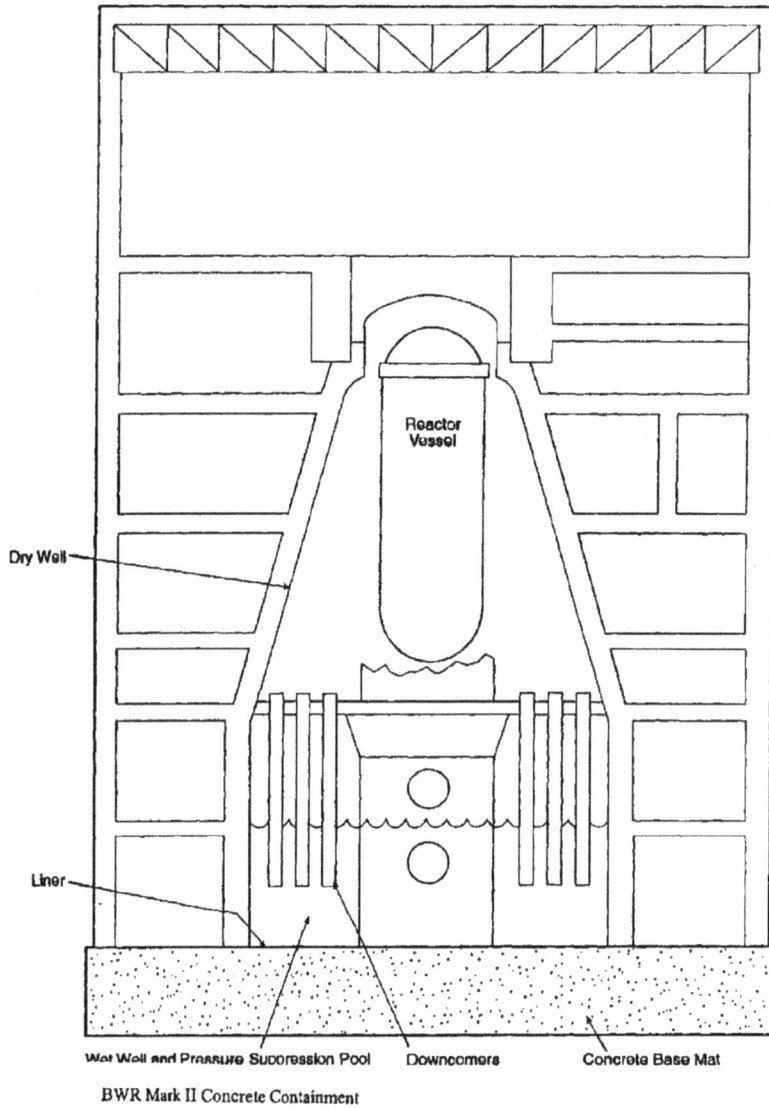

Dry Well

Liner

Reactor Vessel

Wet Well and Pressure Suppression Pool   Downcomers   Concrete Base Mat

BWR Mark II Concrete Containment

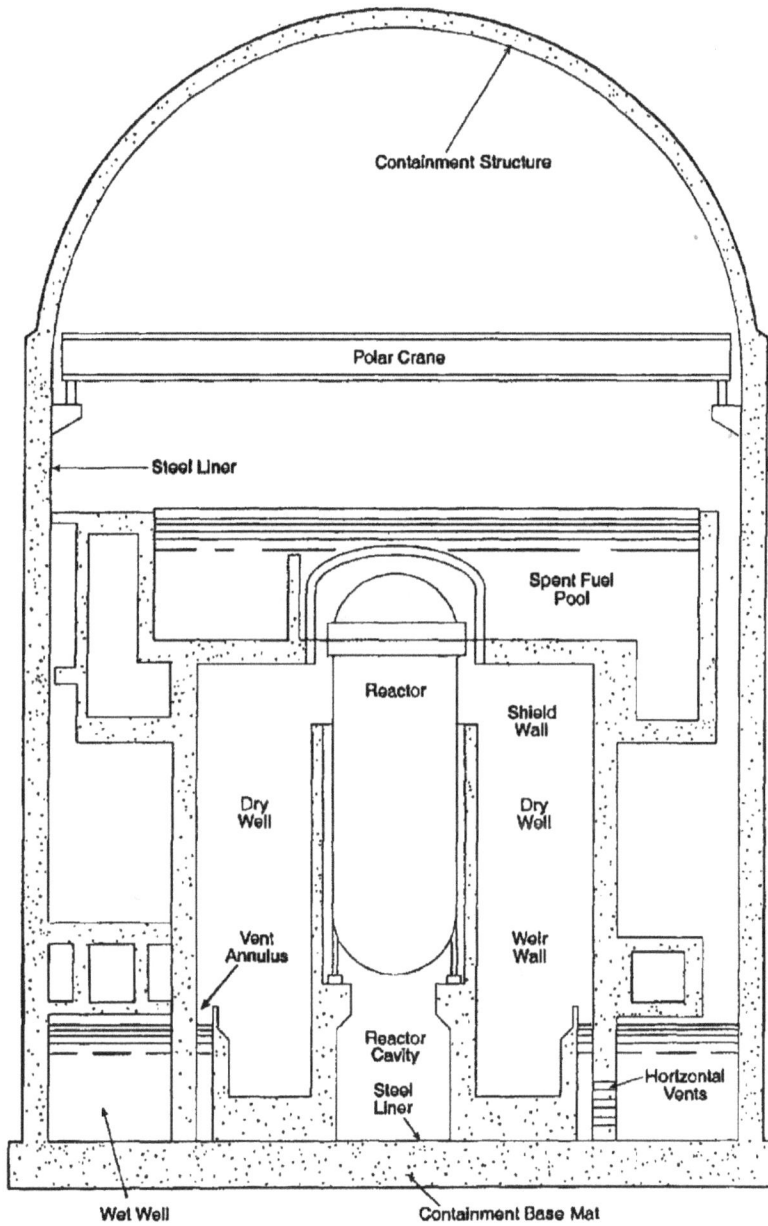

Containment Structure

Polar Crane

Steel Liner

Spent Fuel
Pool

Reactor

Shield
Wall

Dry
Well

Dry
Well

Vent
Annulus

Weir
Wall

Reactor
Cavity

Steel
Liner

Horizontal
Vents

Wet Well

Containment Base Mat

BWR Mark III Concrete Containment

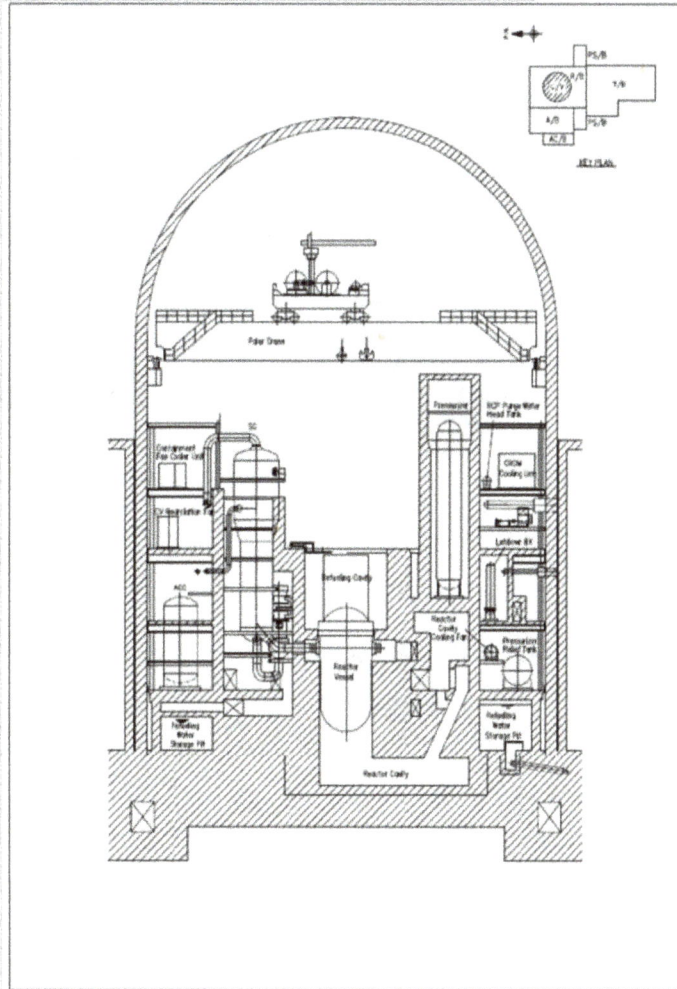

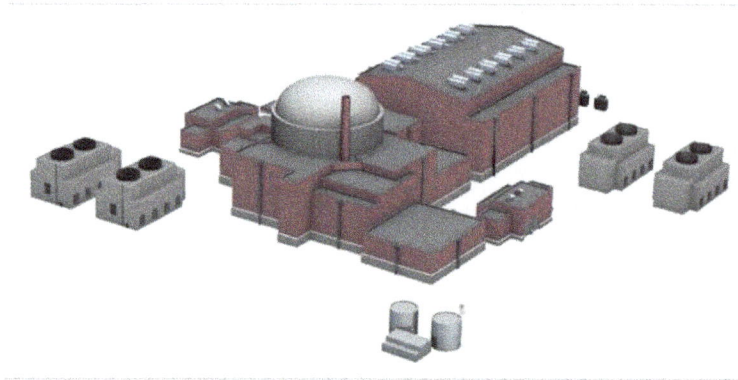

AREVA's U.S. EPR

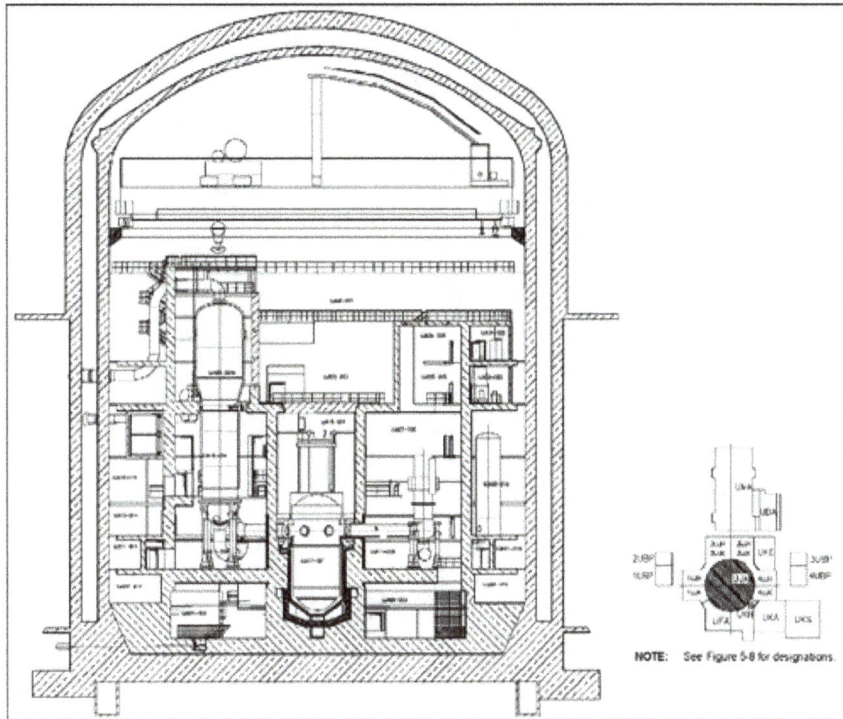

Figure 5-4
Reactor Building – Section A-A

NOTE: See Figure 5-8 for designations.

GE-Hitachi Nuclear Energy

26A6642AJ

Revision 4

September 2007

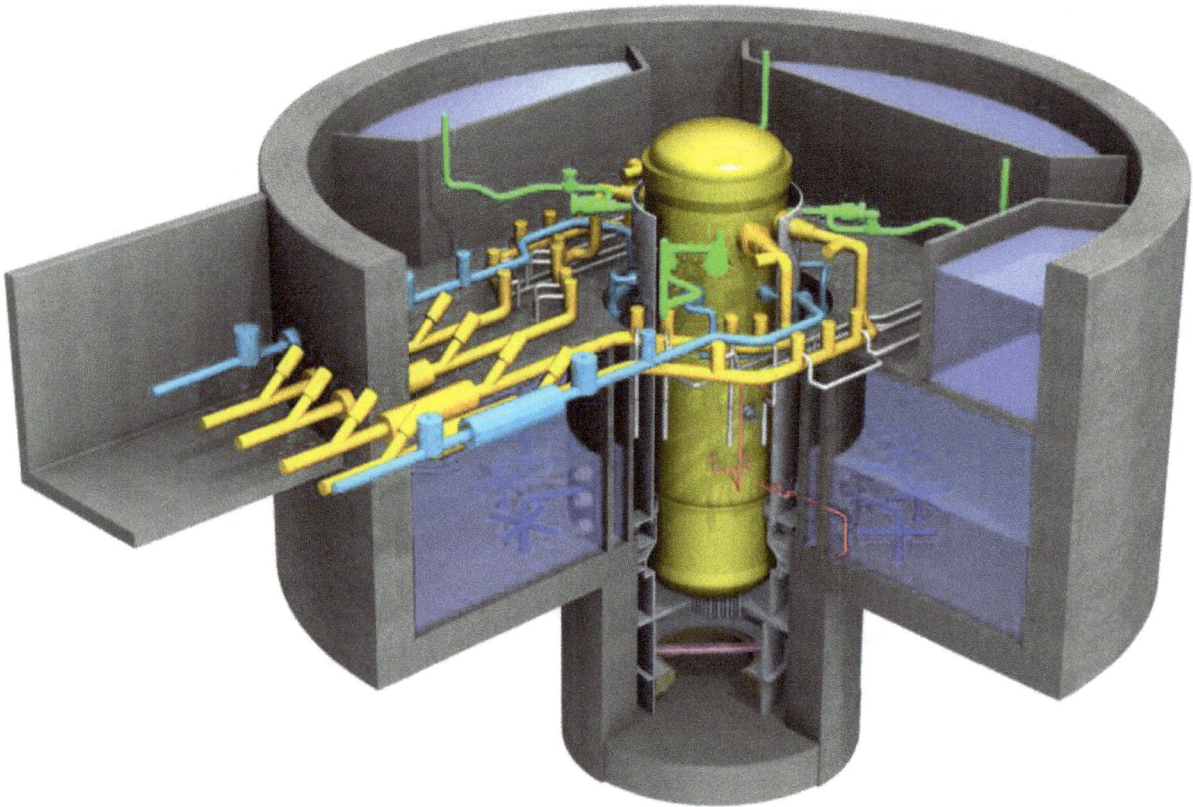

ESBWR Design Control Document

Tier 2

Chapter 3

Design of Structures, Components, Equipment, and Systems

Sections 3.1 - 3.8

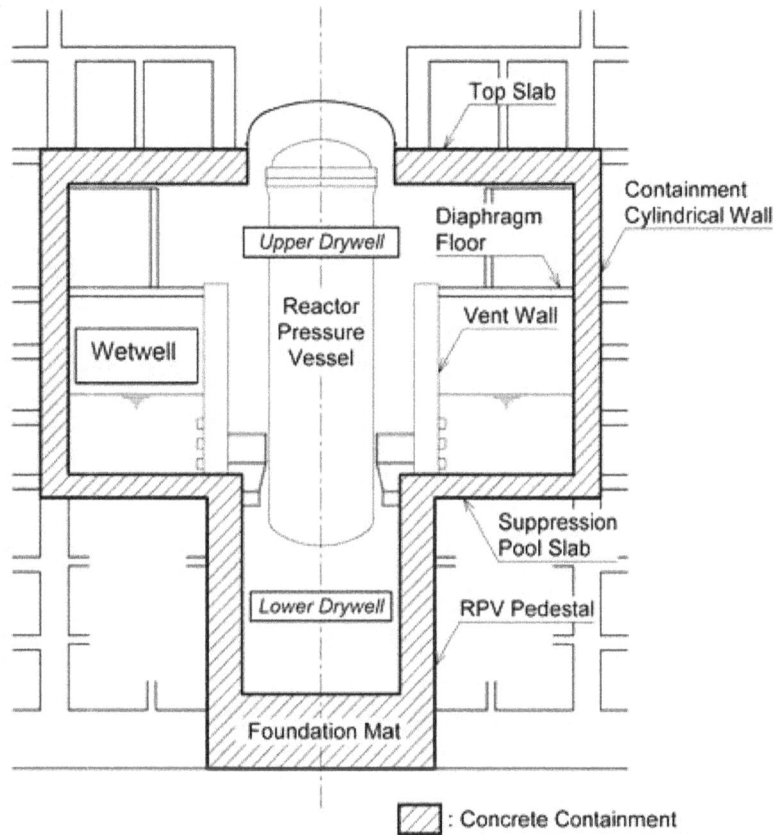

Figure 3.8-1. Configuration of Concrete Containment

APPENDIX B—GENERAL DESIGN AND ANALYSIS INFORMATION

[B-1] FIGURES IN THIS APPENDIX SHOW TYPICAL FINITE ELEMENT MODELS FOR THE OVERALL SHELL, BASE MAT, DISCONTINUITIES AND AN EQUIPMENT HATCH.

EXAMPLE OF THE USE FINITE ELEMENT ANALYSES WHEN DESIGNING PRE-STRESSED CONCRETE REACTOR CONTAINMENTS

The design examples are taken from projects carried out by Scanscot Technology AB during 1992-2008, except for the one shown in Figure 6 that is taken from the paper "Increased plastic strains in containment steel liners due to concrete cracking and discontinuities in the containment structure", P. Anderson and O. Jovall, presented at 19th SMiRT Conference, August 2007.

In the Nuclear Power industry, finite element (FE) models and FE analysis have been used extensively during the years. The capabilities and the range of structures studied, as well as the detailing level of the models, have increased hand in hand with the development of both computer hardware and software. In Figures 1, 2 and 3 respectively, this is exemplified by different generations of FE models of prestressed concrete reactor containments, ranging from axisymmetric models (Figure 1) to wedge models (Figure 2) to fully three-dimensional nonlinear FE models (Figure 3).

Nowadays, not only the containment but also all safety-related buildings at a nuclear facility are on a regular basis modeled and analyzed using the FE method. In Figure 4 FE models of a whole Nuclear Power Plant (NPP) unit including all safety-related buildings is shown, with the containment, the reactor pressure vessel, and the turbine foundation highlighted.

When it comes to the reactor containment, both the global overall behavior, and the local behavior of important areas, can be studied in detail. In Figure 3 a detailed fully three-dimensional global model of a typical PWR containment is shown. The model includes nonlinear description of the concrete (a), the reinforcement (b), all the prestressing tendons modeled as separate units (c), and the steel liner with attached larger penetrations such as the equipment hatch (E/H) and the airlocks (A/L).

The modeling of the prestressing tendons is rather straightforward if the tendons are grouted, i.e. they may not be able to slip. However, even if the tendons are unbonded, i.e. slip is possible to occur between the tendons and the concrete, there exist techniques to model this behavior. In Figure 3, one technique is presented. Using this technique it is also possible in the FE model to directly simulate the prestressing and then automatically calculate the initial losses due to friction and wobbling, slippage during anchoring, and elastic compression of the structure.

The detailed structural response in the concrete structure at major penetrations (E/H and A/L) is studied using sub-models with more refined meshing (Figure 5 (a)). The behavior of the major penetrations themselves (the steel parts) is studied using the same technique (Figure 5 (b)). The steel liner is analyzed using detailed models, here exemplified in Figure 6 by a detailed FE model of the liner in the area connecting to the E/H.

Pipe rupture analyses, including local dynamical effects, can be studied using FE models. In Figure 7(a) the containment structure together with all the piping to be studied is shown, including pipe supports and pipe whip restraints. Using this model, or a subsection of it, the pipe whip behavior during a pipe rupture can be studied (Figure 7 (b)). Also the local dynamical response of impacted concrete structures (Figure 7 (c)) and steel structures (Figure 7 (d)) may be analyzed.

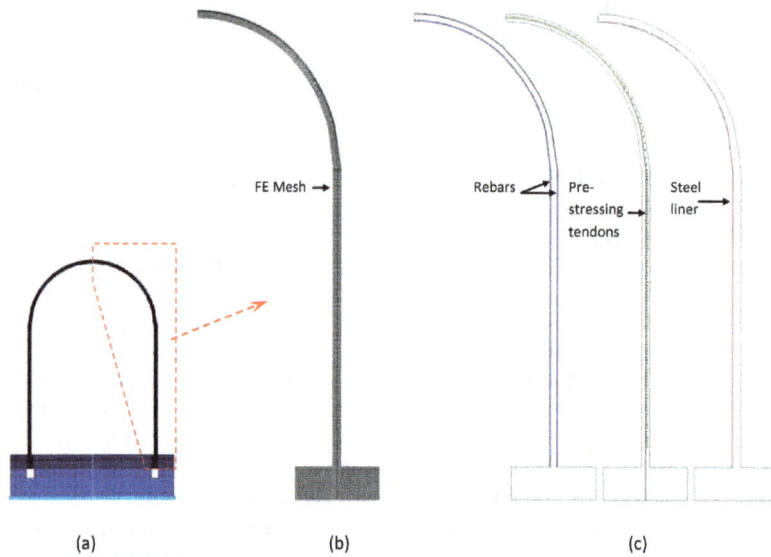

FIGURE 1   AXISYMMETRIC FE MODELING OF PRESTRESSED CONCRETE REACTOR CONTAINMENT (PWR). (A) CONTAINMENT LAYOUT. (B) FE MESH. (C) REBARS, PRESTRESSING TENDONS, AND STEEL LINER.

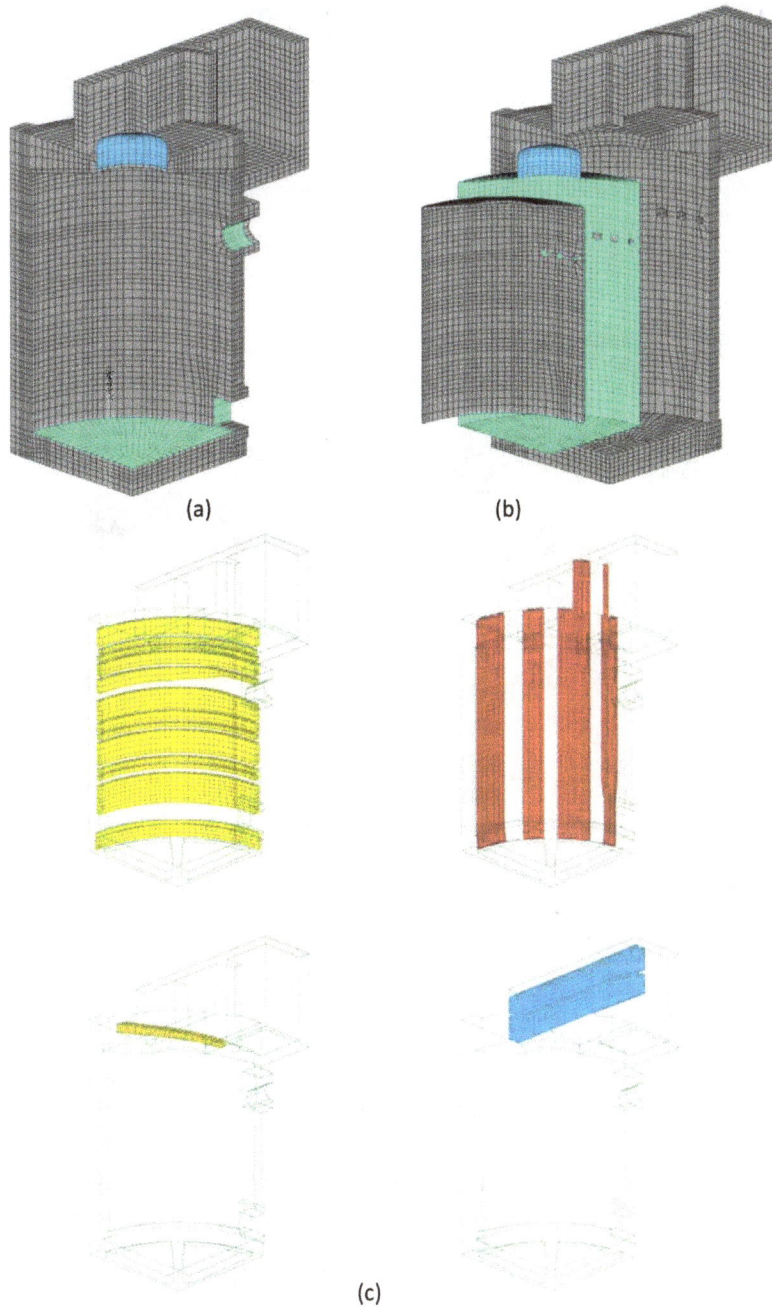

FIGURE 2   (A) WEDGE MODELING OF PRESTRESSED CONCRETE REACTOR CONTAINMENT (BWR). (B) FE MESH (CONCRETE WITH REBARS AND EMBEDDED STEEL LINER). (C) PRESTRESSING TENDONS IN CYLINDRICAL PART (HOOP AND VERTICAL DIRECTION), ROOF AND BASIN WALLS.

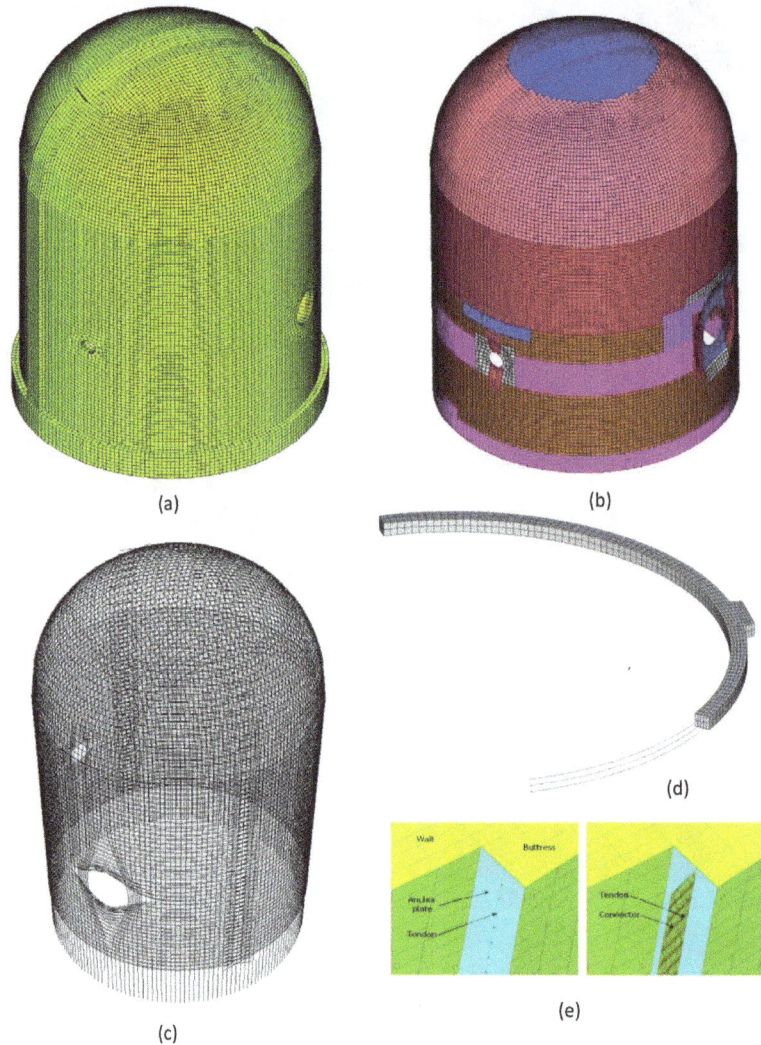

FIGURE 3   FULLY THREE-DIMENSIONAL FE MODELING OF PRESTRESSED CONCRETE REACTOR CONTAINMENT. (A) NONLINEAR CONCRETE WITH STEEL LINER ATTACHED. (B) REBARS. (C) PRESTRESSING TENDONS. (D) TENDONS MODELED WITH TRUSS ELEMENTS, THE INTERACTION BETWEEN TENDONS AND CONCRETE (UNBONDED TENDONS). (E) TENSIONING OF TENDONS IN THE MODEL LEADING TO AUTOMATIC CONSIDERATION OF INITIAL LOSSES (FRICTION, ANCHOR SLIP, AND ELASTIC SHORTENING).

**FIGURE 4   FE MODEL OF ALL SAFETY-RELATED BUILDINGS AT A NPP, INCLUDING THE REACTOR CONTAINMENT
(1), THE PRESSURE VESSEL (2), AND THE TURBINE FOUNDATION (3).**

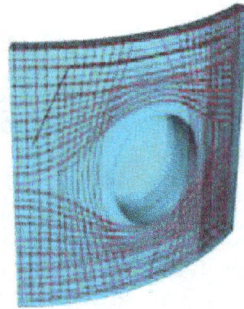

(a)

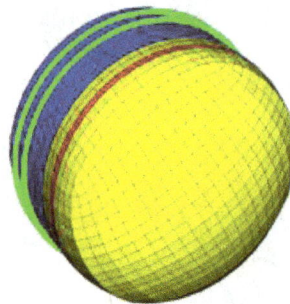

(b)

FIGURE 5 MAJOR PENETRATIONS SUCH AS E/H AND A/L ARE STUDIED USING SUB-MODELS. (A) STUDY OF THE CONCRETE PART SURROUNDING THE PENETRATION INCLUDING REFINED MODELING OF THE CONCRETE. ALSO, THE CASING TUBE AND THE ADJOINING STEEL LINER ARE INCLUDED IN THE MODEL. (B) DETAILED ANALYSIS OF THE PENETRATION ITSELF, INCLUDING ALL STEEL PARTS AND A COMPREHENSIVE MODELING AND DESCRIPTION OF THE INTERACTION WITH THE CONCRETE PART.

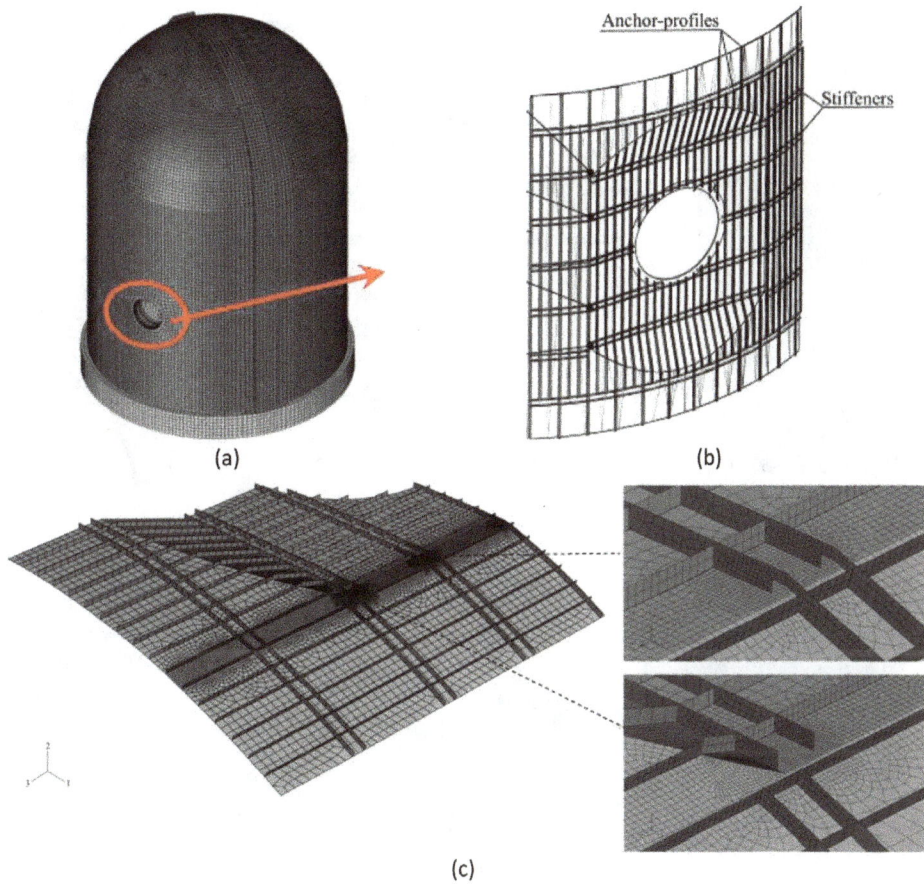

FIGURE 6   DETAILED FE MODEL OF THE STEEL LINER IN THE VICINITY OF THE E/H. (A) CONTAINMENT WITH THE ANALYZED PENETRATION (E/H) HIGHLIGHTED. (B) THE STEEL LINER LAYOUT AT THE E/H. (C) DETAILED QUARTER MODEL OF THE STEEL LINER, THE ANCHOR-PROFILES AND THE STIFFENERS AT THE E/H.

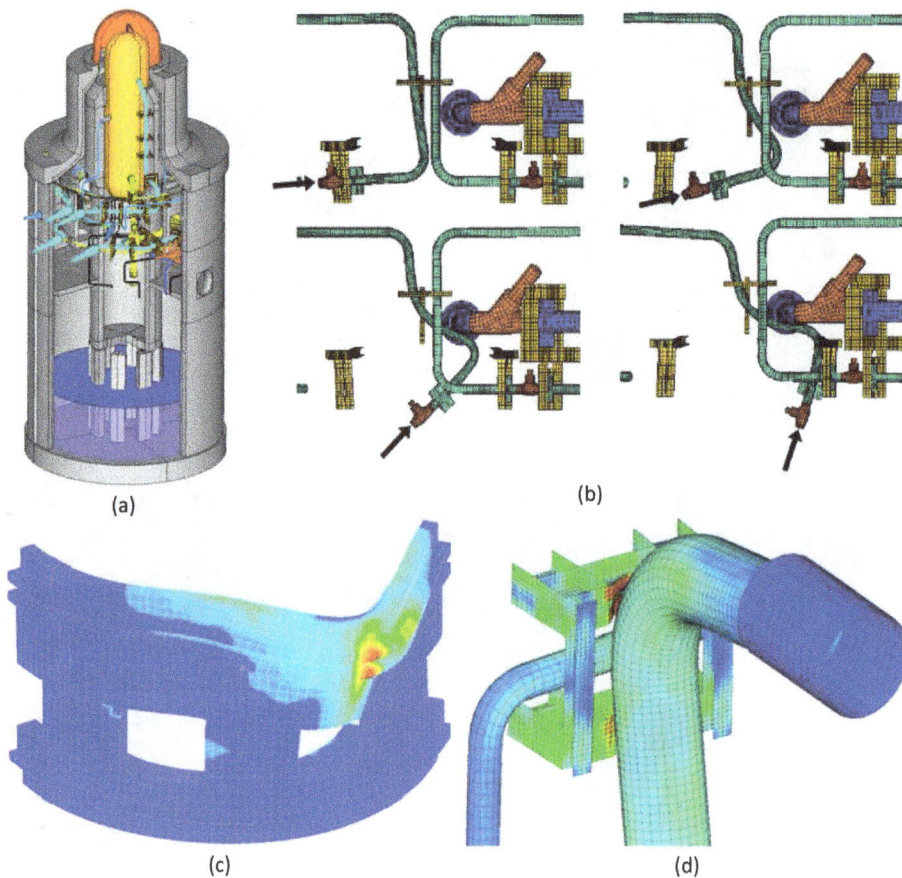

(a)

(b)

(c)

(d)

FIGURE 7   EXTRACT FROM A COMPLETE STUDY OF PIPE RUPTURE AT A PLANT, INCLUDING THE SIMULATION OF PIPE WHIP BEHAVIOR, AND THE STUDY OF LOCAL DYNAMICAL EFFECTS. (A) THE REACTOR CONTAINMENT INCLUDING ALL PIPES OF IMPORTANCE TOGETHER WITH PIPE SUPPORTS AND PIPE WHIP RESTRAINTS. (B) SIMULATION OF PIPE WHIP BEHAVIOR AFTER PIPE RUPTURE. (C) NONLINEAR ANALYSIS OF IMPACTED CONCRETE STRUCTURE. (D) NONLINEAR ANALYSES OF IMPACTED PIPE WHIP RESTRAINT.

PAST FINITE ELEMENT MODELS

Cylinder - Dome Finite Element Model

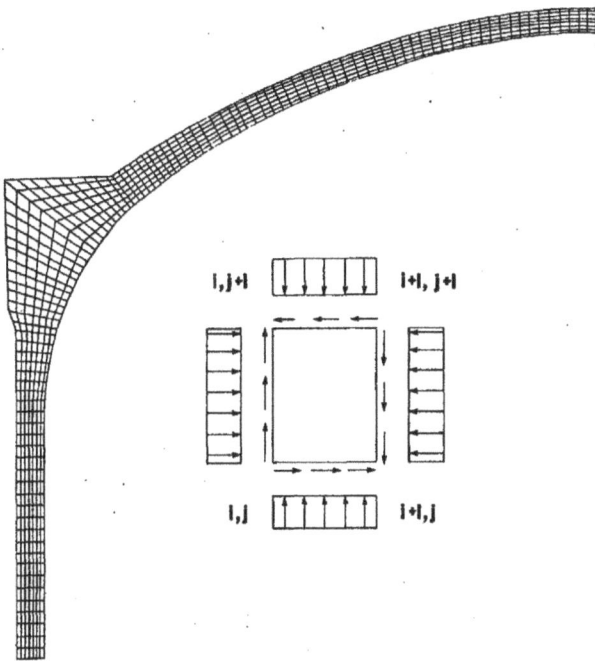

Base Slab - Cylinder Finite Element Model

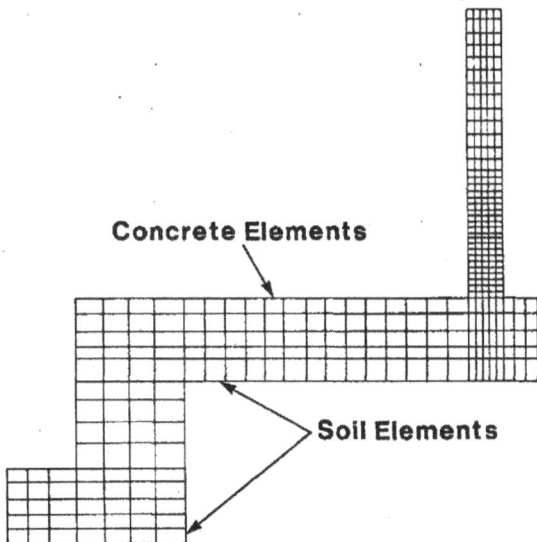

Concrete Elements

Soil Elements

Equipment Hatch Finite Element Model

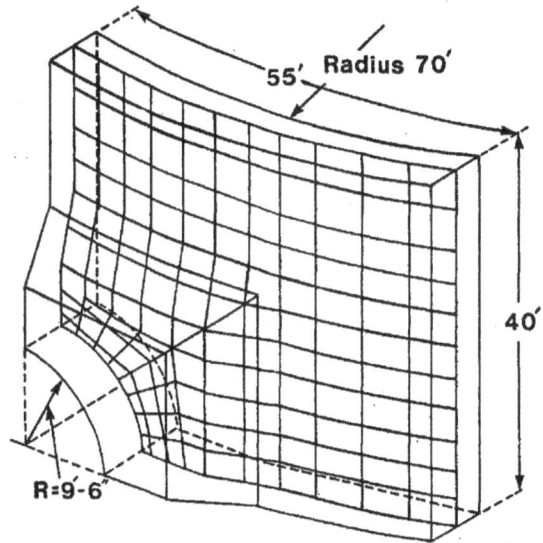

55' Radius 70'

40'

R=9-6'

[B-2] THERMAL EFFECTS IN WALLS OF NUCLEAR CONTAINMENTS—ELASTIC AND INELASTIC BEHAVIOR, G. GURFINKEL, 1ST SMIRT CONFERENCE, 1971.

SUMMARY

This paper was one of the first covering what is typically know as a post-processor of forces and moments that are usually determined by finite computer analysis programs.

Design of containment structures requires evaluation of the thermal effects created by the temperature differential between inside and outside faces. Numerical methods are presented which permit evaluation of thermal effects on reinforced concrete sections subjected to axial load and bending. The wall section of an actual containment is studied for thermal effects under various loading conditions using both elastic and inelastic analysis. The results are compared and interpreted using interaction diagrams.

[B-3] "OPTIMUM DESIGN OF REINFORCED CONCRETE FOR NUCLEAR CONTAINMENTS, INCLUDING THERMAL EFFECTS," KOHLI, T., AND GURBUZ, O., PROCEEDINGS, SECOND ASCE SPECIALTY CONFERENCE ON STRUCTURAL DESIGN OF NUCLEAR PLANT FACILITIES (NEW ORLEANS, 1975), AMERICAN SOCIETY OF CIVIL ENGINEERS, NEW YORK, 1976, V. 1-B. PP. 1292–1319.

[Paper is only available through ASCE]

SUMMARY

The design of reinforced concrete members In accordance with the ACI 318 code is usually a straight-forward task. In the design of containments for nuclear power plants, however, this task is more difficult due to the following reasons:

• the design usually involves more loading combinations,
• different stress and strain limitations are specified under different loading conditions
• consideration of thermal effects is usually required.

The purpose of this paper is to discuss a recently developed computer code for reinforced concrete design for nuclear containments and other concrete structures. In this code, different stress and strain limitations are incorporated. Furthermore, thermal gradients are considered.

[B-4] INCLUDED HERE IS A CLASSICAL SHELL ANALYSIS OF CONTAINMENT DISCONTINUITIES WHICH IS AN EXCERPT FROM "BC-TOP-5A, PRESTRESSED CONCRETE NUCLEAR CONTAINMENT STRUCTURES, H. REUTER, ET AL, FEBRUARY, 1975" BECHTEL POWER CORPORATION.

CAVEAT: THIS MATERIAL REPRODUCED HERE HAS BEEN PREPARED BY AND FOR THE USE OF BECHTEL POWER CORPORATION AND ITS RELATED ENTITIES. ITS USE BY OTHERS IS PERMITTED ONLY ON THE UNDERSTANDING THAT THERE ARE NO REPRESENTATIONS OR WARRANTIES, EXPRESS OR IMPLIED, AS TO THE VALIDITY OF THE INFORMATION, ACCURACY OR CONCLUSIONS CONTAINED HEREIN.

DESIGN EXAMPLE DE-5

Shell Analysis by Classical Methods
Case I – Cylindrical shell at and near the base slab.

Constants –
a = 66.9 ft.
h = 3.83 ft.
E = 590400 K/sq.ft.
$v = 0.17$
$v^2 = 0.03$

Evaluating $\beta = \sqrt[4]{\frac{3(1-v^2)}{a^2 h^2}} = 0.0816$

$$D = \frac{Eh^3}{12(1-v^2)} = 2849600$$

The solution of the classical differential equation for a semi-infinite cylindrical shell is, by Timoshenko (see reference 1)

$$w = e^{-\beta x}(C_3 \cos\beta x + C_4 \sin\beta x) + f(x) \quad (1)$$

Where w is the radial displacement (positive inward) and $f(x)$ is the particular solution dependant on the loading. In this case there is pressure loading and $f(x)$ is constant.

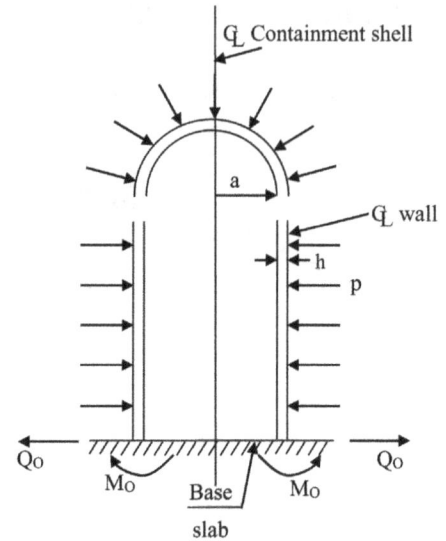

Taking successive derivatives and defining MO and QO as the edge loading at x = 0, the constants C3 and C4 are eliminated by noting that:

$$-D\frac{d^2w}{dx^2}(at\ x=0)M_O \quad \text{and} \quad -D\frac{d^3w}{dx^3}(at\ x=0)=Q_O$$

CAVEAT: THIS REPORT REPRODUCED HERE HAS BEEN PREPARED BY AND FOR THE USE OF BECHTEL CORPORATION AND ITS RELATED ENTITIES. ITS USE BY OTHERS IS PERMITTED ONLY ON THE UNDERSTANDING THAT THERE NO REPRESENTATIONS OR WARRANTIES, EXPRESS OR IMPLIED, AS TO THE VALIDITY OF THE INFORMATION, ACCURACY OR CONCLUSIONS CONTAINED HEREIN.

This results in the following equations –

$$w = -\frac{e^{-\beta x}}{2\beta^3 D}[\beta M_O(\cos\beta x - \sin\beta x) + Q_O\cos\beta x] + f(x) \quad (2)$$

$$\theta = \frac{dw}{dx} = \frac{e^{-\beta x}}{2\beta^2 D}[2\beta M_o\cos\beta x + Q_o(\cos\beta x + \sin\beta x)] + f'(x) \quad (3)$$

$$M_x = -D\frac{d^2w}{dx^2} = \frac{e^{-\beta x}}{\beta}[\beta M_O(\cos\beta x + \sin\beta x) + Q_O\sin\beta x] - Df''(x) \quad (4)$$

$$Q_x = -D\frac{d^3w}{dx^3} = e^{-\beta x}[Q_o(\cos\beta x - \sin\beta x) - 2\beta M_o\sin\beta x] - Df'''(x) \quad (5)$$

NOTE: $f(x) = f(x) = f(x) = 0$ for a constant pressure loading. Also, $f(x)$ becomes the free inward radial deformation, which is constant, due to pressure.

From equations 2) and 3), using full restraint at x=0,

$$w(at\ x=0) = -\frac{1}{2\beta^3 D}[\beta M_O + Q_O] + f(x) = 0 \quad (6)$$

$$\theta(at\ x=0) = \frac{1}{2\beta^2 D}[2\beta M_O + Q_O] = 0 \qquad (7)$$

where $f(x) = \frac{pa^2}{Eh}\left(1 - \frac{v}{2}\right)$ solving 6 and 7

$$M_O = \frac{2-v}{12}\sqrt{\frac{3}{(1-v^2)}}\,hap$$

$$Q_O = \frac{2-v}{6}\sqrt{\frac{3}{1-v^2}}\,\beta hap$$

For a pressure of p = −8.64 K/sq.ft. = −60 psi,
M_O and Q_O become −
M_O = 593.4 ft.K/ft. and Q_O = −98.3 k/ft.

Case II – Hemispherical Head-Cylinder Interface
Constant values:
Subscript 1 refers to cylinder.
Subscript 2 refers to head.

$$\beta_1 = \sqrt[4]{\frac{3(1-v^2)}{a^2 h_1^2}} = 0.0816$$

$$D_1 = \frac{Eh_1^3}{12(1-v^2)} = 2849600$$

$$*\beta_2 = \sqrt[4]{\frac{3(1-v^2)}{a^2 h_2^2}} = 0.0949$$

$$*D_2 = \frac{Eh_2^3}{12(1-v^2)} = 1149600$$

where E = 590400 k/sq. ft.
v = .17

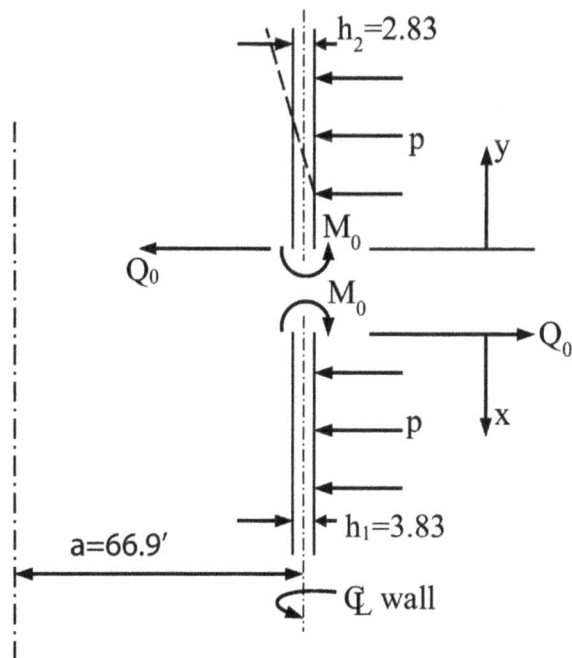

*NOTE: For stiffness considerations, a cylinder is used to simu-
late the head. The free cylinder and head increases in radius due to
pressure are:

$$\delta_1 = -\left(1 - \frac{v}{2}\right)\frac{a^2}{Eh_1}p \qquad \delta_2 = -\frac{1-v}{2}\frac{a^2}{Eh_2}p$$

Let δ = Final inward radial displacement at interface (x=0)
θ = Final rotation at interface (+ is counter-clockwise)

$$\delta = \frac{1}{2D_1\beta_1^2}\left[-\frac{Q_O}{\beta_1} - M_O\right] + \frac{pa^2}{Eh_1}\left(1 - \frac{v}{2}\right) \qquad (8)$$

$$\delta = \frac{1}{2D_2\beta_2^2}\left[\frac{Q_O}{\beta_2} - M_O\right] + \frac{pa^2}{2Eh_2}(1-v) \qquad (9)$$

$$\theta = \frac{1}{2D_1\beta_1^2}[Q_O + 2\beta_1 M_O] \qquad (10)$$

$$\theta = \frac{1}{2D_2\beta_2^2}[Q_O - 2\beta_2 M_O] \qquad (11)$$

Equating (8) to (9) and (10) to (11), and letting

$$G_1 = \frac{\beta_1}{\beta_2} = \sqrt{\frac{h_2}{h_1}} \qquad\qquad G_2 = \frac{D_1}{D_2} = \left(\frac{h_1}{h_2}\right)^3$$

$$K_1 = 2\beta_1\left[\frac{G_1 G_2 + 1}{G_1^2 G_2 - 1}\right] \qquad K_2 = \frac{a^2}{E}(D_2\beta_2^2)\left[\frac{2-v}{h_1} - \frac{1-v}{h_2}\right]$$

$$K_3 = \frac{K_1}{\beta_2} - 1 + \left[\frac{K_1}{\beta_1} + 1\right]\left[\frac{1}{G_1^2 G_2}\right] \quad K_4 = \frac{K_2}{K_3}$$

$$K_5 = (K_1)(K_4)$$

Then:

$$M_0 = K_4 p \qquad \text{and} \qquad Q_0 = K_5 p$$

For this case:

$h_1 = 3.83$	$v = .17$	$a = 66.9$
$h_2 = 2.83$	$\beta_1 = 0.0816$	$D_2 = 1149600$
$E = 590400$	$\beta_2 = 0.0949$	$p = -8.64$

From which,

$G_1 = .8596$	$G_2 = 2.4788$	$K_1 = .6144$
$K_2 = 14.480$	$K_3 = 10.1313$	$K_4 = 1.4292$
$K_5 = 0.8781$		

$$M_0 = K_4 p = -(1.4292)(8.64) = -12.35 * \text{ft.K/ft.}$$

$$Q_0 = K_5 p = -(0.8781)(8.64) = -7.59 \text{ K/ft.}$$

*NOTE: This value (−12.35) is not the greatest value of the
moment. The greatest value is (−35.9) at a point 8 ft. down the
cylinder from the interface.

The resulting generalized equations for the head become:

$$M_y = \frac{e^{-\beta_2 Y}}{\beta_2}[-12.35\beta_2(\mathrm{Cos}\,\beta_2 Y + \mathrm{Sin}\,\beta_2 Y) \qquad (12)$$

$$+7.59\,\mathrm{Sin}\,\beta_2 Y]$$

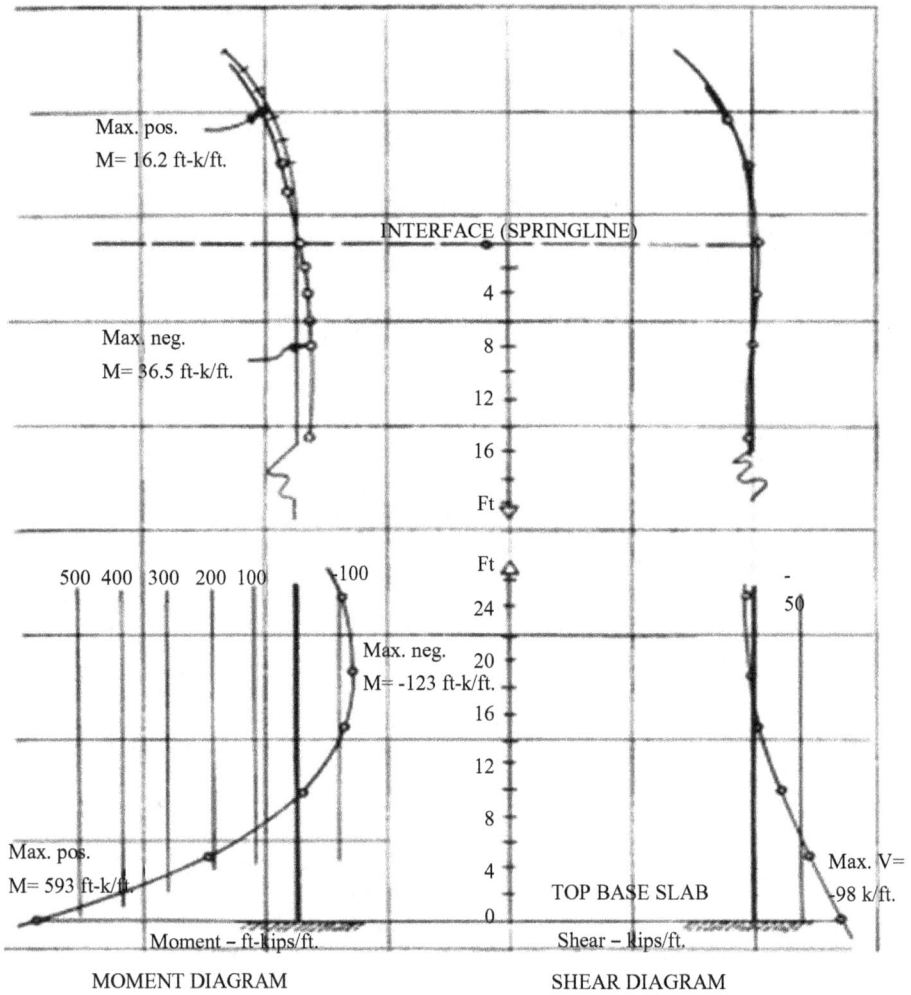

MOMENT DIAGRAM   SHEAR DIAGRAM

FIGURE DE5-1

$$Q_y = e^{-\beta_z Y}[7.59(\mathrm{Sin}\beta_2 Y - \mathrm{Cos}\beta_2 Y) - (\beta_2 x 24.70\,\mathrm{Sin}\beta_2 Y)] \quad (13)$$

The resulting generalized equations for the cylinder immediately below the head become:

$$M_x = \frac{e^{-\beta_1 x}}{\beta_1}[-12.35\beta_1(\mathrm{Cos}\beta_1 x + \mathrm{Sin}\beta_1 x) - 7.59\,\mathrm{Sin}\beta_1 x] \quad (14)$$

$$Q_x = e^{-\beta_1 x}[7.59(\mathrm{Sin}\,\beta_1 x - \mathrm{Cos}\beta_1 x) + (\beta_1 x 24.70\,\mathrm{Sin}\beta_1 x)] \quad (15)$$

Results from Equations (4), (5), (12), (13), (14), and (15) are shown in Figure DE5-1.

APPENDIX C—EXAMPLES RADIAL SHEAR AND RADIAL TENSION TIE

[C-1] RADIAL SHEAR TIE EXAMPLE

These are excerpts from "BC-TOP-5A, PRESTRESSED CONCRETE NUCLEAR CONTAINMENT STRUCTURES, H. Reuter et al, FEBRUARY 1975" Bechtel Power Corporation

[C-2] RADIAL TENSION TIE EXAMPLE

These are excerpts from "BC-TOP-5A, PRESTRESSED CONCRETE NUCLEAR CONTAINMENT STRUCTURES, H. Reuter et al, FEBRUARY 1975" Bechtel Power Corporation

DESIGN EXAMPLE DE-2

Radial Tie Reinforcement for Radial Shear

Significant radial shear in the containment shell occurs at the junction of cylindrical wall to base slab and the transition between hemispherical dome and wall or ring girder to dome and wall. The cylindrical wall at penetrations is also designed for radial shear due to pipe rupture forces and the reaction from design pressure and temperature.

The following example illustrates the design techniques given in Appendix C, Section CC-3411.4.2. The loading values used in the example are summarized in Table DE2-1 for the junction of a containment cylinder to the base slab. In the following example the moments and shears caused by vertical membrane forces have been neglected.

Given Information:
(a) Loading values in Table DE-2-1
(b) $d = 38''$, $y_t = 21''$, $b' = 12''$

(c) $A_S = 2.00$ in^2 (vertical bonded steel)
(d) $f_c' = 5000$ psi
(e) $f_y = 60,000$ psi

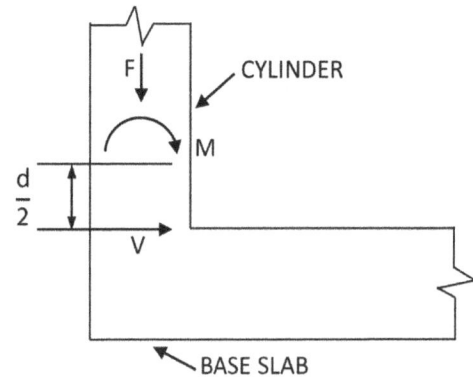

Evaluate constants:

$$\sqrt{f_c'} = 70.7 \qquad n = \frac{505}{\sqrt{f_c'}} = 7.15$$

$$\rho = \frac{2.00}{(12)(38)} = .0044 \qquad n\rho = .0314$$

$$K = 1.75 - .36/n\rho + 4.0n\rho = 0.73 \text{ (note: } K \geq .6 \text{ for } \rho \geq .003)$$

$$K\sqrt{f_c'} = 52 \text{ psi} \qquad \frac{I}{y_t} = 3520 \text{ in}^3$$

$$6\sqrt{f_c'} = 425 \text{ psi} \qquad 3.5\sqrt{f_c'} = 248 \text{ psi}$$

Sign Convention:
(+) Moment: compression on inside face
(+) Force: compression

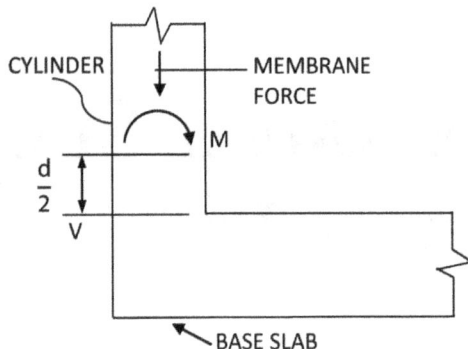

CASE I – SERVICE LOAD – DEAD LOAD PLUS ONE HALF HOOP PRESTRESSING AT TRANSFER OF PRESTRESS.

For Equation 6 and Equation 7, refer to Appendix C.

$$V_i = 0; \qquad M_i = 0; \qquad V = \frac{106}{2} = 53 \text{ k/ft and}$$

$$M\frac{422}{2} = 211 \text{ k-ft/ft}$$

$$f_{pe} = f_{pc} = 220 \text{ psi (compression)}$$

$$M_{CR} = \frac{3520}{12000}(425 + 220) = 190 \text{ k-ft/ft}$$

Equation 7 $\qquad V_{ci} = 52 + \dfrac{\left[(190)\dfrac{53}{211} - 0\right]1000}{12(38)} = 156 \text{ psi}$

Equation 6 $\qquad V_{cw} = 248\sqrt{1 + \dfrac{220}{248}} = 340 \text{ psi}$

Since this is a service load, the allowable is reduced in accordance with Section CC-3421.3.

$$v_c' = .50(156) = 78 \text{ psi}$$

$$v = \frac{V}{b'd} = \frac{(53)(1000)}{12(38)} = 117 \text{ psi}$$

Reinforce for (117-78) = 39 psi or minimum requirements, whichever is greater, using service load allowables.

TABLE DE2-1   LOADINGS THE VALUES PROVIDED IN THIS TABLE ARE FOR ILLUSTRATION PURPOSES ONLY.

Type of Load	Membrane Forces (k/ft)		Membrane Stress (psi) Vert.	Moment (k-ft/ft)	Shear (k/ft)
	Hoop	Vert.			
Dead Load (D)	—	110	220	—	—
Prestress (F)					
(a) Effective	643	322	644	350	87
(b) At Transfer	773	462	924	422	106
(c) Ultimate	915	486	972	500	125
Pressure (P_a)					
(a) P_a	−550	−275	−550	−300	− 75
(b) P_t=1.15P_a	−630	−316	−632	−345	− 86
(c) 1.25P_a	−690	−385	−770	−375	− 87
(d) 1.50P_a	−825	−412	−824	−450	−105
Temperature (T)					
(a) T_t	− 40	− 40	− 80	− 22	− 6
(b) T_o	− 50	− 50	−100	− 28	− 7
(c) T_a	−100	−100	−200	− 55	−14
Earthquake (E_o)					
1.0E_o					
(a) Horizontal	—	−160	−320	—	—
(b) Vertical	—	− 15	− 30	—	—
		−175	−350*		
1.25E_o					
(a) Horizontal	—	−200	−400	—	—
(b) Vertical	—	− 19	− 38	—	—
		−219	−438*		

*For simplicity, the seismic values are combined by absolute summation. (Actually, they shall be combined by the SRSS method.)

CASE II – SERVICE LOAD – DEAD LOAD PLUS FULL HOOP AND VERTICAL PRESTRESS WITH THE FOLLOWING SEQUENCE:

1st – Dead Load

2nd – One Half Hoop-Prestress

3rd – Vertical Prestress

4th – Final One Half Hoop-Prestress

$$V_i = 53 \text{ k/ft}; \qquad M_i = 211 \text{ k-ft/ft}; \qquad V = 53 \text{ k/ft}$$

$$M = 211 \text{ k/ft}$$

$$f_{pe} = f_{pc} = 220 + 924 = 1144 \text{ psi (compression)}$$

$$M_{CR} = \frac{3520}{12000}(425 + 1144) = 460 \text{ k-ft/ft}$$

Equation 7 $\qquad V_{ci} = 52 + \dfrac{\left[(460-211)\dfrac{53}{211} + 53\right]1,000}{(12)(38)} = 306 \text{ psi}$

Equation 6 $\qquad V_{cw} = 248\sqrt{1 + \dfrac{1144}{248}} = 588 \text{ psi}$

Reduction due to service load:

$$V_c' = .50(306) = 153 \text{ psi}$$

$$V = \frac{(106)(1000)}{12(38)} = 233 \text{ psi}$$

Reinforce for $(233 - 153) = 80 \text{ psi}$

CASE III – SERVICE LOAD – TEST CONDITION

$$(1.0D + 1.0F + 1.0T_t + 1.0P_t)$$

(Note: Effective prestress used)

$$V_i = -87 \text{ k/ft}; \qquad M_i = -350 \text{ k-ft/ft}; \qquad V = 92 \text{ k/ft}$$

$$M = 367 \text{ k/ft}$$

$$f_{pe} = f_{pc} = (220 + 644 - 80 - 632) = 152 \text{ psi (compression)}$$

$$M_{CR} = \frac{3,520}{12,000}(425 + 152) = 169 \text{ k-ft/ft}$$

Equation 7 $\quad V_{ci} = 52 + \dfrac{\left[(169+350)\dfrac{92}{367} - 87\right]1000}{12(38)} = 147 \text{ psi}$

Equation 8 $\qquad V_{cw} = 248\sqrt{1 + \dfrac{152}{248}} = 315 \text{ psi}$

Reduction due to service load:

$$v_c' = .50(147) = 73 \text{ psi}$$

$$v = \frac{(92-87)1,000}{12(38)} = 11 \text{ psi}$$

Only minimum reinforcement required.

CASE IV – SERVICE LOAD – NORMAL CONDITION

$$(1.0D + 1.0F + 1.0T_o + 1.0E)$$

$$V_i = -87 \text{ k/ft}; \qquad M_i = -350 \text{ k-ft/ft}; \qquad V = 7 \text{ k/ft}$$

$$M = 28 \text{ k-ft/ft}$$

$$f_{pe} = f_{pc} = (220 + 644 - 100 - 350) = 414 \text{ psi (compression)}$$

$$M_{CR} = \frac{3520}{12000}(425 + 414) = 245 \text{ k-ft/ft}$$

Equation 7 $\quad v_{ci} = 52 + \dfrac{\left[(245+350)\dfrac{7}{28} - 87\right]1,000}{(12)(38)} = 187 \text{ psi}$

Equation 6 $\qquad v_{cw} = 248\sqrt{1 + \dfrac{414}{248}} = 406 \text{ psi}$

Reduction due to service load:

$$v_c' = .50(187) = 93 \text{ psi}$$

$$v = \frac{(87-7)1,000}{12(38)} = 176 \text{ psi}$$

Reinforce for $(176 - 93) = 83 \text{ psi}$

CASE V – FACTORED LOAD – ABNORMAL CONDITION

$$(1.0D + 1.0F + 1.0T_a + 1.5P_a)$$

$$V_i = -87 \text{ k/ft}; \qquad M_i = -350 \text{ k-ft/ft}; \qquad V = 119 \text{ k/ft}$$

$$M = 505 \text{ k-ft/ft}$$

$$f_{pe} = f_{pc} = (644 + 220 - 824 - 200) = -160 \text{ psi (tension)}$$

$$M_{CR} = \frac{3,520}{12,000}(425 - 160) = 78 \text{ k-ft/ft}$$

Equation 7 $\quad v_{ci} = 52 + \dfrac{\left[(78+350)\dfrac{119}{505} - 87\right]1{,}000}{12(39)} = 83 \text{ psi}$

Equation 6 $\quad v_{cw} = 248\sqrt{1 - \dfrac{160}{248}} = 147 \text{ psi}$

Since f_{pc} is tensile but less than $3.5\sqrt{f'_c}$, then check Equation 5. (Refer to Appendix C.)

$$v_c = 2.0\sqrt{5{,}000}[1 - (.002)(160)] = 96 \text{ psi} > v_{ci}$$

$$v = \frac{(119-87)1{,}000}{(.85)(12)(38)} = 83 \text{ psi}$$

Since $v - v_c = 0$, use minimum reinforcing requirements.

CASE VI – FACTORED LOAD – ABNORMAL/ SEVERE ENVIRONMENTAL CONDITION

$$(1.0D + 1.0F + 1.0T_a + 1.25P_a + 1.25E_o)$$

$$V_i = -87 \text{ k/ft}; \qquad M_i = -350 \text{ k-ft/ft}; \qquad V = 101 \text{ k/ft}$$

$$M = 430 \text{ k-ft/ft}$$

$$f_{pe} = f_{pc} = (644 + 220 - 200 - 438 - 770) = -544 \text{ psi (tension)}$$

Since $f_p = -544 \text{ psi} > 3.5\sqrt{f'_c}$, reinforcing shall be provided to resist

$$v = \frac{(101-87)1000}{.85(12)(38)} = 36 \text{ psi}$$

(a) Shear Reinforcement Requirement

Load Case IV controls. Note that if diagonal ties were to be used, they would not be effective in resisting the shear of Load Case VI. Use horizontal ties as follows:

Assume ties spaced 18″ c.c. vertically and 12″ c.c horizontally. Since this is a Service Load condition,

$$A_v = \frac{(v_u - v_C)}{.5f_y}bs$$

$$= \frac{83 \times 12 \times 18}{30{,}000} = .598 \text{ in}^2/\text{ft. (use \# 7 ties)}$$

(b) Check for Minimum Shear Reinforcement

(Refer to Sec. 4.5.8.1 of this Topical Report)

$$A_v = \frac{(50)bs}{.5f_y} = \frac{50 \times 12 \times 18}{30{,}000} = .36 \text{ in}^2/\text{ft.} < .598 \text{ in}^2/\text{ft}$$

Provide reinforcement requirement based on (a).

RADIAL TENSION TIES

(Ref. to Sec. 4.5.9)

Hemispherical Dome

Two regions are examined to determine the required area of reinforcing steel ties to be provided to resist the tensile forces created in the concrete by the curvature of the tendons. There are: (1) the spherical triangular region where both vertical inverted U-shaped tendon groups and the dome hoop tendons are superimposed, and, (2) the region where only the inverted U-shaped tendon groups are located. See Figure DE2-1 for schematic tendon arrangement.

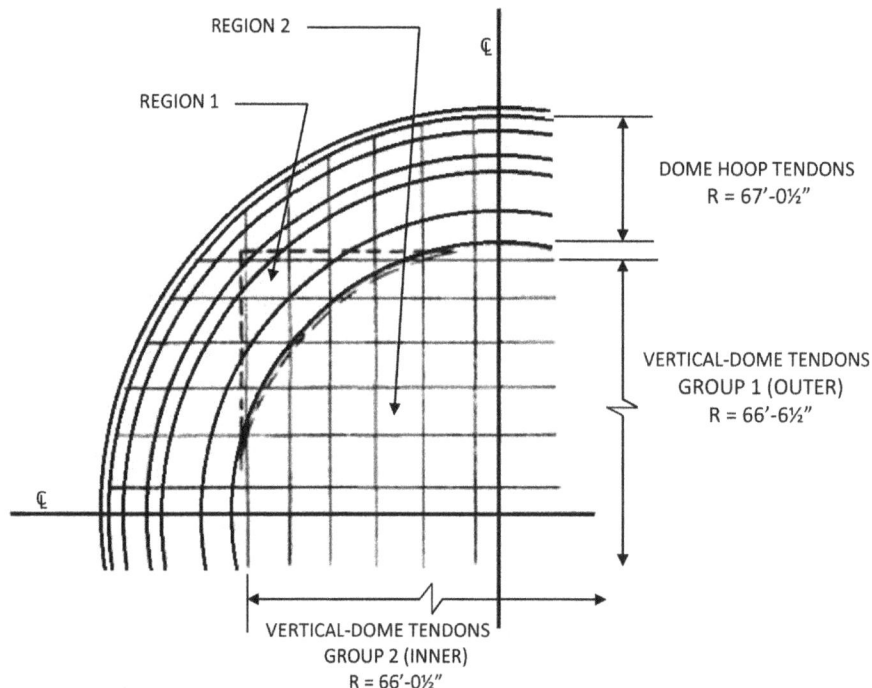

FIGURE DE2-1   SCHEMATIC PLAN OF DOME TENDONS (ONE QUADRANT SHOWN)

If there were no friction during stressing, the radial component of prestressing force from each tendon would be uniform throughout the dome. Since friction does exist, the maximum radial component occurs where the tendon stress is greatest. For region (1) this is about at the springline, while for region (2) the maximum occurs at the edge of the 90° solid angle around the center line of the dome and at 45° to the planes of the dome tendons. In this example, the dimensions and prestressing forces are the same as in DE-1.

Region 1:

Required vertical prestress = 337 k/ft, effective. Required dome hoop prestress = 337 k/ft, effective.

Vertical tendon component:

Average radius = 1/2 (66.542 + 66.042) = 66.292 ft.

Vertical tendon radial force = $\dfrac{F}{R} = \dfrac{174.12 \times 337}{107.5 \times 66.292} = 8.23$ k/ft^2.
(at transfer)

Hoop tendon component:

Radius = 67.042 ft.

Dome tendon radial force = $\dfrac{F}{R} = \dfrac{161.5 \times 337}{127.0 \times 67.042} = 6.39$ k/ft^2.
(at transfer)

Total radial force, Fr, = 6.39 + 2(8.23) = 22.85 k/ft^2.

Criterion (A):

Assume dome thickness = 2'-11"

Radius to outside surface of dome = 65.000 + 2.917 = 67.917 ft.

Radius to centroid of tendons = 66.542 ft.

Distance from centroid to surface = 1.375 ft.

$$A_v = \frac{1.375}{2.917} \times \frac{Fr}{(.5f_y)} = \frac{1.375}{2.917} \times \frac{22.85}{(.5)(60)} = .359 \text{ in}^2/\text{ft}.$$

Criterion (B):

$$A_v = \frac{8.23}{(.5)f_y} = .274 \text{ in}^2/\text{ft}^2$$

Criterion (A) controls; use #6 hairpin radial ties.

$$A_v = 2(.44) = .88 \text{ in}^2.$$

$$\text{Area covered} = \frac{.88}{.359} = 2.45 \text{ ft}^2.$$

The spacing of the #6 hairpins is determined by each project based on the area covered (2.45 ft^2) and the layout of the typical dome reinforcement.

Region 2:

Requred vertical prestress = 337 k/ft., effective.

Vertical tendon component: average radius = 66.292 ft.

Vertical tendon radial force (at transfer) =

$$\frac{F}{R} = \frac{(174.1 + 135.4)(337)}{2(107.5)(66.292)} = 7.318 \text{ k/ft}^2.$$

Total radial force, Fr = 2(7.318) = 14.636 k/ft^2.

Criterion (A):

Assume dome thickness = 2'-11"

Radius to outside surface of dome = 67.917 ft.

Radius to centroid of tendons = 66.292 ft.

Distance from centroid to surface = 1.625 ft.

$$A_v = \frac{1.625}{2.917} \times \frac{Fr}{(.5f_y)} = \frac{1.625}{2.917} \times \frac{14.636}{(0.5)(60)} = .272 \text{ in}^2/\text{ft}^2.$$

Criterion (B):

$$A_v = \frac{7.318}{(.5((f_y))} = .244 \text{ in}^2/\text{ft}^2.$$

Criterion (A) controls; use #6 hairpin radial ties.

$$A_v = .88 \text{ in}^2$$

$$\text{Area covered} = \frac{.88}{272} = 3.24 \text{ ft}^2.$$

The spacing of the #6 hairpins is determined by each project based on the area covered (3.24 ft^2) and the layout of the typical dome reinforcement.

Cylindrical Wall:

Required hoop prestress = 674 k/ft.

Hoop tendon component:

Radius = 67.300 ft.

$$Fr = \frac{161.5}{127} \times \frac{674}{67.300} = 12.74 \text{ k/ft}^2. \text{ (at transfer)}$$

Criterion (A):

Assume containment wall thickness = 3'-10"

Radius to outside surface of cylinder = 65.00 + 3.833 = 68.833 ft.

Radius to centroid of hoop tendons   = 67.300 ft.

Distance from centroid to surface   = 1.533 ft.

$$A_v = \frac{1.533}{3.833} \times \frac{12.74}{(.5)(60)} = .170 \text{ in}^2/\text{ft}^2$$

(Note: Criterion (B) is not applicable to cylinder walls since there is only one layer of curved tendons.)
Use #6 hairpin radial ties.

$$A_v = 2(.44) = .88 \text{ in}^2.$$

$$\text{Area covered} = \frac{.88}{.170} = 5.18 \text{ ft}^2.$$

The spacing of the #6 hairpins is determined by each project based on the area covered (5.18 ft^2) and the layout of the typical wall reinforcement.
(Note: In areas where reinforcement is required for shear and radial tension, the area of reinforcement furnished is the larger calculated for either shear or radial tension.)
R is the radius to the spherical surface that contains the center lines of the tendons of each group.

APPENDIX D—DETAILED SHELL DESIGN

[D-1] PRESTRESSED CONCRETE SHELL DESIGN. EXCERPTS FROM "BC-TOP-5A, PRESTRESSED CONCRETE NUCLEAR CONTAINMENT STRUCTURES, H. REUTER, ET AL., FEBRUARY 1975", BECHTEL POWER CORPORATION

CAVEAT: THIS MATERIAL REPRODUCED HERE HAS BEEN PREPARED BY AND FOR THE USE OF BECHTEL POWER CORPORATION AND ITS RELATED ENTITIES. ITS USE BY OTHERS IS PERMITTED ONLY ON THE UNDERSTANDING THAT THERE ARE NO REPRESENTATIONS OR WARRANTIES, EXPRESS OR IMPLIED, AS TO THE VALIDITY OF THE INFORMATION, ACCURACY OR CONCLUSIONS CONTAINED HEREIN.

1. INTRODUCTION

This design example is presented to show the methods employed and the factors considered to establish the level of prestress for a prestressed concrete containment. Calculations are also given which are performed to assure that structural integrity is maintained. It must be pointed out that this example does not analyze all possible material properties, or all load combinations and locations in the structure which must be investigated. Calculations here are samples only and individual containments should be thoroughly checked by the designer.

This design example is based on the containment configuration and material properties given below:

Containment Data:

$P_a = 60$ psig

$E_S = 29 \times 10^6$ psi

$E_S = 4.1 \times 10^6$ psi

Creep and shrinkage losses $= 500 \times 10^{-6}$ in/in

$T_O = 100°F$ (operating temperature at inside face)

$T_a = 320°F$ (accident temperature at inside face)

$R = 65$ ft

$D_1 \cong 0$ k/ft

$D_2 = 28.2$ k/ft

$D_3 = 116.7$ k/ft

$f'_c = 6000$ psi

$f_y = 60,000$ psi

where:
D_1, D_2, and D_3 are Dead Load at points shown in Figure D-1

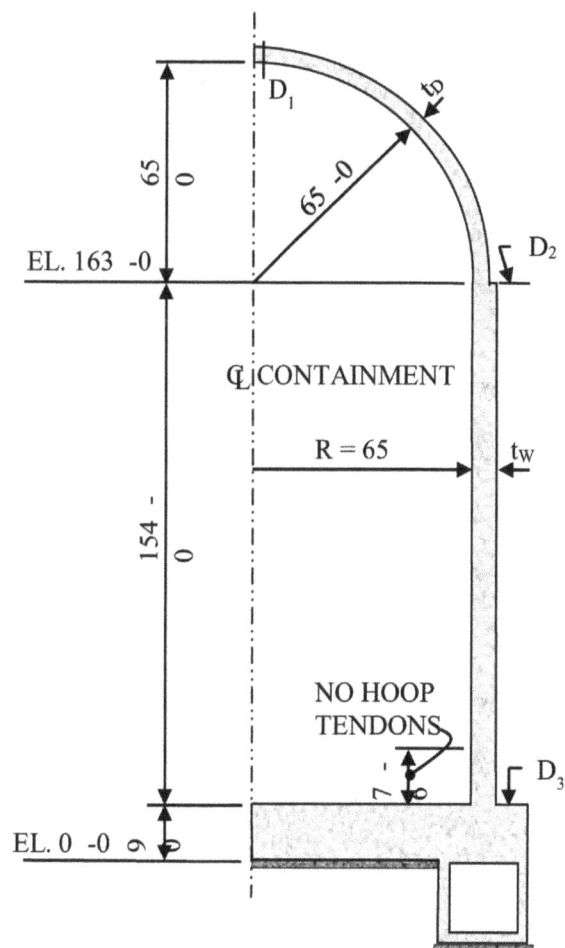

FIGURE D-1

Before proceeding with the design, approximate thickness for the various portions of the shell must be chosen. These may be taken from past experience and, in this example, are 3'–6″ for the wall and 2'–6″ for the dome. These dimensions are assumed, at the beginning, to be the minimum section depth required so that the allowable stresses in Appendix C are satisfied. After these dimensions are shown to be adequate in this design example, they would be used in all further analysis such as finite element analysis of the shell.

Additional section thickness must be added to allow for variations during construction. Present radial tolerances are ±3″ for the wall liner plate erection and ±1″ for form placement. Thus, 4 inches must be added to the typical wall section. The increased thickness should then be used in the seismic analysis.

2. SEISMIC DATA

A dynamic analysis, using the average section thickness described above, is performed to obtain equivalent static forces in the structure as discussed in BC-TOP-4. The results of this analysis are usually similar to Figures 7-22 and 7-23 in this topical and consist of two perpendicular horizontal earthquake components and one vertical earthquake component. The effects of these three components must be considered simultaneously.

The resultant forces due to seismic loads are determined using the SRSS combination of the stresses from each seismic component as given in Article CC-3521.1.1a of Appendix C of this topical report.

An example of how these forces are obtained is given below:

$$N_{vhx} = \frac{E_{mx}(X)t}{I} = \frac{E_{mx}}{\pi R^2} \cos\phi$$

$$N_{vhy} = \frac{E_{my}(X)t}{I} = \frac{E_{my}}{\pi R^2} \cos\phi$$

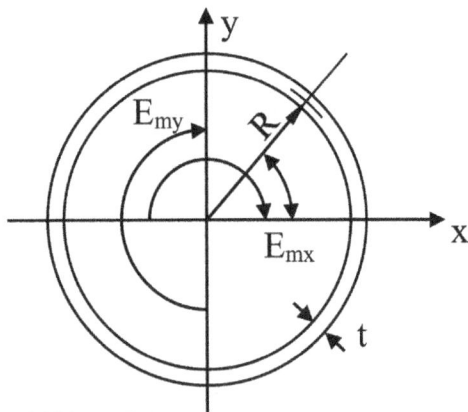

FIGURE D-2

In the above, and are moments due to the horizontal components which vary with elevation in the structure.

$$V_{uhx} = \frac{E_{sx}(Q)}{2It} = \frac{E_{sx}}{\pi R}\sin\phi$$

$$V_{uhy} = \frac{E_{sy}(Q)}{2It} = \frac{E_{sy}}{\pi R}\sin\phi$$

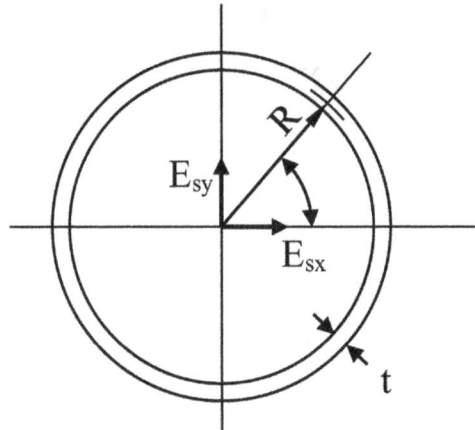

FIGURE D-3

In the above, E_{sx} and E_{sy} are shears due to the horizontal seismic components which vary with elevation in the structure.

E_z is the vertical force due to the vertical seismic component which varies with elevation in the structure.

$$N_{vv} = \frac{E_z}{A} = \frac{E_z}{2\pi r}$$

The following Table D-1 gives sample values for seismic forces, shears, and moments of an Operating Basis Earthquake. Tables D-2 and D-3 give the shear and normal forces for various locations in the containment wall. The combined shear () and normal forces () are calculated by the following equations:

$$V_u = [(V_{uhx})^2 + (V_{uhy})^2]^{1/2}$$

$$N_{ve} = [(N_{vhx})^2 + (N_{vhy})^2 + (N_{vv})^2]^{1/2}$$

3. LEVEL OF PRESTRESS

The level of prestress is defined as: $X = \dfrac{F+D}{P_a}$
To assure structural integrity, $X \geq 1.2$.

The tangential shear design criteria in Article CC-3521.1.1a of Appendix C of this topical report requires that bonded reinforcing steel be provided if sufficient membrane compression is not available to counteract shear and membrane loads due to seismic. If reinforcing is required to resist these forces or any other membrane load, Article CC-3531 of the design criteria requires that it be mechanically spliced if there are tensile stresses in the concrete perpendicular to the reinforcement.

In the meridional direction, if $X = 1.2$ is supplied at the apex of the dome, sufficient normal compression will usually exist in the wall so that reinforcing is not required for resisting membrane tension and tangential shear force. In the hoop direction, the dead load membrane compression does not exist and therefore, if only $X = 1.2$ is provided, reinforcing steel may be required to resist tangential shear. Hoop prestressing may be increased so that reinforcing is not required to resist tangential shear and membrane loads in the hoop direction.

The decision of whether to increase the level of prestress or use mechanical splices in the reinforcement is left to the project. Consideration should be given to the decrease in tendon spacing and the increase in the required section thickness due to increase in level of prestress.

This design example assumes a 1.2 level of prestress.

4. SAMPLE PRESTRESS CALCULATIONS

Given:

$f'_c = 6000$ psi

$f_{Pu} = 240$ ksi

$\mu = 0.14$

$K = 0.0003$

Long term losses:

Creep and shrinkage $= 500 \times 10^{-6}$ in/in

Tendon relaxation $= 8\%$ at 70 f_{pu}

Tendon sheathing diameter $= 6$ in

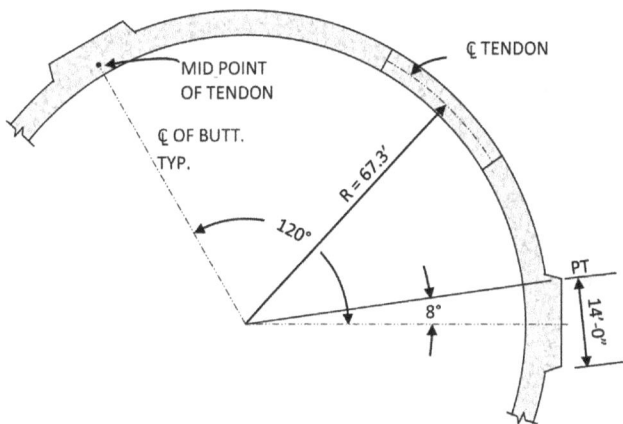

FIGURE D-4

4.1 Containment Wall – Hoop Prestressing

$$F = 1.2\, PR$$

$$= 1.2\frac{(60)(65)(144)}{1000} = 674 \text{ k/ft}$$

4.2 Minimum Tendon Stress Due To Friction

$$f_{px} = f_{ps}\, e^{-(\mu\alpha + KL)}$$

$$\alpha = (120 - 8) \times \frac{\pi}{180} = 1.955 \text{ Rad.}$$

$$L = 14 + 67.3 \times 1.955 = 145.6 \text{ ft.}$$

$$\mu\alpha + KL = 0.14 \times 1.955 + 0.0003 \times 145.6 = 0.31738$$

$$f_{ps} = 0.8\, f_{pu} = 192.0 \text{ ksi (maximum jacking stress)}$$

$$f_{px} = f_{ps}\, e^{-(.3174)} = (192)(0.728) = 139.8 \text{ ksi (at midpoint of tendon)}$$

4.3 Tendon Stress and Location Of Point A Due to Friction

$$f_{ppA} = \frac{0.8f_{pu} + 0.7f_{pu}}{2} = 0.75f_{pu} = 180 \text{ ksi}$$

$$e^{(\mu\alpha + KL)} = \frac{f_{ps}}{f_{pa}} = \frac{192}{180} = 1.067$$

$$(\mu\alpha + KL) = 0.065$$

$$0.14\alpha + 0.0003(14 + 67.3\alpha) = 0.065$$

$$0.160\alpha = 0.065 - 0.0042 = 0.0608$$

$$\alpha = \frac{0.0608}{0.160} = 0.38 \text{Rad.} = 21.77°$$

4.4 Average Stress In Tendon Due To Friction

$$f_{avg} = \frac{\text{Area under the curve (at transfer) in Figure D-5}}{\text{Applicable Length}}$$

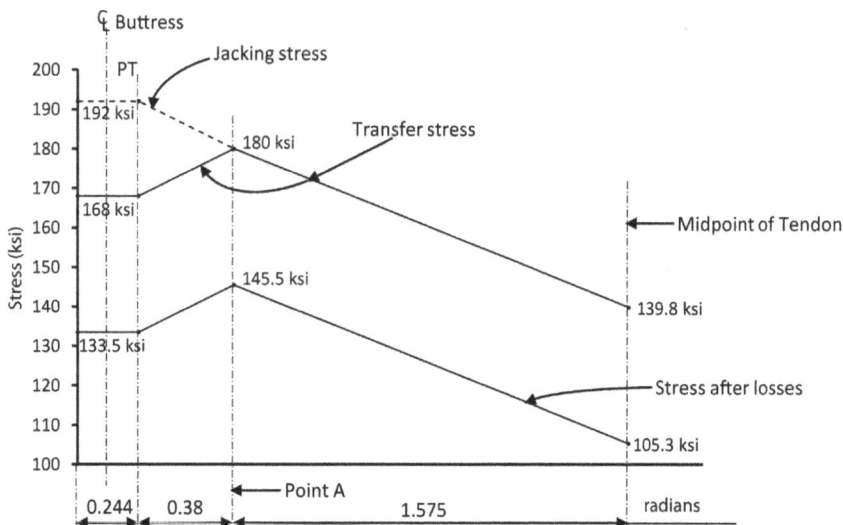

FIGURE D-5

$$f_{avg} = 0.68\,f_{pu} = 0.68 \times 240$$

$$f_{avg} = 163.2 \text{ ksi}$$

4.5 Other Losses

a. Elastic shortening

$$\frac{f_c}{2}\left(\frac{E_s}{E_c}\right) = \frac{.30f_c'}{2}\left(\frac{E_s}{E_c}\right) = \frac{1.8}{2}\left(\frac{29 \times 10^6}{4.1 \times 10^2}\right) = 6.6 \text{ ksi}$$

where f_c is assumed to be $.30f_c'$

b. Creep and shrinkage

$$500 \times 10^{-6} \times 29 \times 10^3 \text{ ksi}$$

c. Relaxation

$$0.08 \times 0.7 \times 240 = 13.4 \text{ ksi}$$

Total losses = $(6.6 + 14.5 + 13.4) = 34.5$ ksi

4.6 Initial Average Tendon Stress Considering Elastic Loss ($f_{E.L.}$) And Average Effective Tendon Stress After All Losses (f_{se})

$$f_{se} = 163.2 - 34.5 = 128.7 \text{ ksi}$$

$$f_{E.L.} = 163.2 - 6.6 = 156.6 \text{ ksi}$$

4.7 Tendon Requirements

Assume $186 - 1/4'' \; \phi$ wire tendons. Note: Three hoop tendons form two complete rings.

$$A_{ps} = 9.13 \text{in}^2$$

$$\text{No. of tendons} = \frac{674(154 - 7.5)}{(2/3)(9.13)(128.7)} + 3 \cong 129 \text{ tendons. (use 129 tendons)}$$

The above number includes three additional tendons for tendon surveillance.

$$\text{Spacing} = \frac{(154 - 7.5)}{129} = 1.14 \text{ ft} = 13.6 \text{ in}$$

4.8 Containment Meridional Prestressing

At apex of dome

$$F = 1.2\frac{PR}{2} = 1.2\frac{(60)(65)(144)}{2 \times 1000} = 337 \text{ k/ft}$$

4.9 Dome And Wall Tendon Stresses After Friction Losses

	Anchor	Springline	Dome at ℄
L (ft)	0.0	163	267*
α (radian)	0.0	0	1.5707
$e^{-(\mu\alpha + KL)}$	1.0	0.9524	0.7407
Max. jacking stress (ksi)	192	182	142.0
Stress @ transfer (ksi)	168	176.2	142.0

*The length based on average radius for the two dome layers.

4.10 Other Losses

a. Elastic shortening = (assumed)	= 6.6 ksi
b. Creep and shrinkage = $500 \times 10^{-6} \times 29 \times 10^3$	= 14.5 ksi
c. Relaxation = 0.08(0.7)(240)	= 13.4 ksi
Total	= 34.5 ksi

4.11 Effective Stress After All Losses (f_{se})

	Anchor	Springline	Dome at ℄
Stress at transfer (ksi)	168	176.2	142.0
Losses (ksi)	34.5	34.5	34.5
Stress after losses (f_{se}) (ksi)	133.5	141.7	107.5

4.12 Tendon Requirements

Assume $186 - 1/4'' \; \phi$ wires.

$$A_{ps} = 9.13 \text{ in}^2$$

$$N = \frac{2\pi(65)(337)}{107.5 \times 9.13} + 4 \cong 144 \text{ tendons (use 72} \cap \text{ tendons)}$$

TABLE D-1

Node		Elevation (ft)	Moment (k-ft)		Shear (k)		Axial (k)
			E_{mx}	E_{my}	E_{sx}	E_{sy}	E_z
Springline	1	163.0	186,310	183,103	6589	6477	3473
	2	141.0	265,327	260,779	8267	8128	4576
	3	119.0	451,140	443,468	10,113	9940	5995
	4	97.0	678,270	666,825	11,629	11,438	7388
	5	75.0	939,085	923,375	12,831	12,623	8750
	6	53.0	1,226,290	1,205,930	13,745	13,524	10,075
	7	31.0	1,533,035	1,507,770	14,402	14,170	11,357
	8	9.0	1,853,075	1,822,745	14,677	14,441	11,357
Base Slab	9	0.0	2,026,470	1,993,410	14,677	14,441	11,357

TABLE D-2   SHEAR FORCES IN CONTAINMENT WALL DUE TO 1.0 TIMES THE OBE (K/FT)

E-W Shear – V_{uhx}

Node/ϕ	0	15	30	45	60	75	90
1	0	8.64	15.72	22.16	27.12	30.24	31.28
2	0	10.84	19.72	27.80	34.00	37.92	39.24
3	0	13.24	24.12	34.00	41.60	46.40	48.00
4	0	15.24	27.72	39.08	47.84	53.32	55.20
5	0	16.80	30.60	43.12	52.72	58.84	60.92
6	0	18.00	32.76	46.20	56.52	63.04	65.28
7	0	18.88	34.36	48.40	59.24	66.04	68.40
8	0	19.24	35.00	49.32	60.36	67.32	69.68
9	0	19.24	35.00	49.32	60.36	67.32	69.68

N-S Shear – V_{uhy}

Node/ϕ	0	15	30	45	60	75	90
1	30.76	29.72	26.64	21.76	15.44	8.48	0
2	38.60	37.28	33.44	27.32	19.40	10.64	0
3	47.20	45.60	40.92	33.40	23.72	13.04	0
4	54.32	52.48	47.04	38.44	27.28	15.00	0
5	59.92	57.88	51.92	42.40	30.12	16.52	0
6	64.20	62.04	55.64	45.44	32.24	17.72	0
7	67.28	65.00	58.28	47.60	33.80	18.56	0
8	68.56	66.24	59.40	48.52	34.44	18.92	0
9	68.56	66.24	59.40	48.52	34.44	18.92	0

Combined Shear Forces – V_u

Node/ϕ	0	15	30	45	60	75	90
1	30.76	30.92	30.92	31.04	31.20	31.40	31.28
2	38.60	38.80	38.80	38.96	39.12	39.40	39.24
3	47.20	47.48	47.48	47.68	47.88	48.16	48.00
4	54.32	54.64	54.60	54.80	55.04	55.40	55.20
5	59.92	60.27	60.28	60.48	60.76	61.12	60.92
6	64.20	64.60	64.56	64.80	65.08	65.48	65.28
7	67.28	67.68	67.64	67.88	68.20	68.60	68.40
8	68.56	68.96	68.92	69.20	69.48	69.92	69.68
9	68.56	68.96	68.92	69.20	69.48	69.92	69.68

The above number includes two additional $\cap$ for tendon surveillance.

$$\text{Spacing} = \frac{2\pi(67.3)}{144(12)} = 35.30 \text{ in}$$

4.13   Dome Hoop Tendons

$$F = 1.2 * \frac{PR}{2} = 1.2 * \frac{(60 \times 65)}{(2 \times 1000)}(144) = 337 \text{ k/ft}$$

4.14   Stresses in the Dome Hoop Tendons

Assume the hoop tendon stresses in the dome are the same as hoop tendon stresses in the cylinder. (See 4.5)

4.15   Tendon Requirements

$$\text{No. of tendons} = \frac{\pi R}{4} \times \frac{F}{2/3(A_{ps} \times f_{se})}$$

$$= \frac{\pi(66.5)}{4} \times \frac{337}{2/3(9.13 \times 128.7)} \cong 23 \text{ tendons (use 24 tendons)}$$

$$\text{Spacing} = \frac{(\pi/4)R}{24} = \frac{(\pi/4)(66.5)}{24}(12) = 26.11 \text{ in}$$

5.   MINIMUM SHELL THICKNESS

5.2   Containment Wall Thickness (Based On Hoop Prestress)

$$F_{provided} = \frac{129}{126}(674) = 690 \text{ k/ft}$$

t_m = minimum wall thickness = ① + ② + t_{net}

① = Thickness (diameter) for sheathing parallel to section

TABLE D-3 FORCES IN CONTAINMENT WALL DUE TO 1.0 TIMES THE OBE (K/FT)

E-W Moment – N_{vhx}

Node/ϕ	0	15	30	45	60	75	90
1	13.20	12.76	11.44	9.32	6.60	3.40	0
2	18.80	18.16	16.28	13.28	9.40	4.88	0
3	31.96	30.88	27.68	22.60	16.00	8.28	0
4	48.04	46.40	41.60	33.96	24.04	12.44	0
5	65.52	64.28	57.60	47.04	33.28	17.20	0
6	86.88	83.92	75.24	61.44	43.44	22.48	0
7	108.60	104.92	94.04	76.80	54.24	28.12	0
8	131.28	126.80	113.68	92.84	65.64	33.96	0
9	143.56	138.68	124.32	101.52	71.80	37.16	0

N-S Moment – N_{vhy}

Node/ϕ	0	15	30	45	60	75	90
1	0	3.36	6.48	9.16	11.24	12.52	12.96
2	0	4.80	9.24	13.08	16.00	17.84	18.48
3	0	8.12	15.72	22.20	27.20	30.36	31.40
4	0	12.24	23.64	33.40	40.92	45.64	47.24
5	0	16.92	32.72	46.24	56.64	63.20	65.40
6	0	22.12	42.72	60.40	74.00	82.52	85.44
7	0	27.64	53.40	75.52	92.52	103.16	106.80
8	0	33.44	64.56	91.32	111.84	124.72	129.12
9	0	36.56	70.60	99.88	122.32	136.40	141.24

Vertical Normal Force – N_{vv} (k/ft)

Node/ϕ	0
1	8.24
2	10.88
3	14.26
4	17.56
5	20.80
6	23.92
7	26.96
8	26.96
9	26.96

Combined Normal Forces – N_{ve}

Node/ϕ	0	15	30	45	60	75	90
1	15.56	15.56	15.52	15.58	15.44	15.40	15.36
2	21.72	21.68	21.64	21.56	21.52	21.44	21.44
3	35.00	34.96	34.88	34.76	34.60	34.52	34.48
4	51.16	51.12	50.96	50.76	50.60	50.44	50.40
5	69.72	69.64	69.44	69.16	68.92	68.72	68.64
6	90.12	90.04	89.76	89.44	89.08	88.80	88.72
7	111.92	111.80	111.48	111.04	110.60	110.28	110.16
8	134.04	133.88	133.52	132.96	132.44	132.08	131.92
9	146.08	145.92	145.52	144.92	144.36	143.92	143.76

② = Equivalent thickness for sheathing perpendicular to section

$$t_{net} = \frac{f_{EL}}{f_{se}} \times \frac{F_{provided}}{f_{c(allow)} \times b}$$

where b = 12″

$$f_{c(allow)} = .35(f'_{ci}) = .35(6000) = 2.1 \text{ ksi}$$

$$t_m = 6 + \frac{\pi(3)^2(2)}{3(13.3)} + \frac{156.8}{128.7} \times \frac{690}{2.1 \times 12}$$

$$t_m = 6 + 1.42 + 33.35 = 40.8 \text{ in}$$

Add 4″ for construction tolerance; wall thickness = 40.8 + 4 = 44.8 in

use $t_w = 3' - 10''$

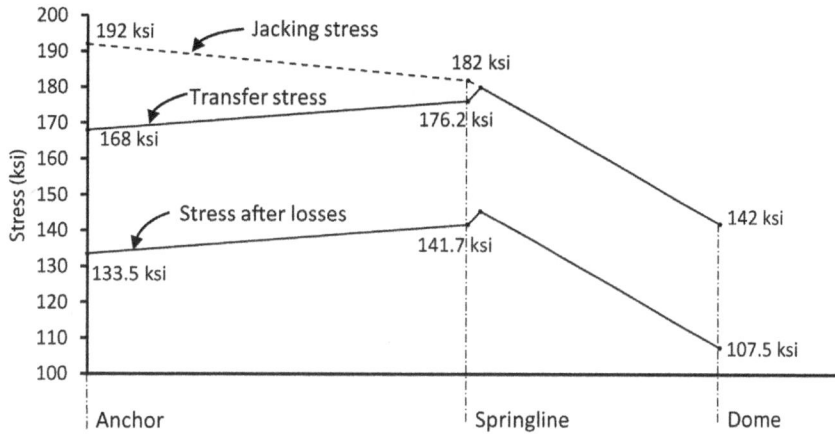

FIGURE D-6

5.2 Dome Thickness At Springline

$$F_{provided} = \frac{144}{140} \times 337 = 346.6 \text{ k/ft}$$

t_m = Minimum wall thickness = ① + ② + t_{net}

① = Thickness (diameter) for sheathing parallel to section
② = Equivalent thickness for sheathing perpendicular to section

$$t_{net} = \left[\left(\frac{f_{EL} \times f_{provided}}{f_{se}}\right) + D_2\right]\frac{1}{f_{c(allow)} \times b}$$

where
    b = 12 in
    D_2 = 28.2 k/ft
    f_{EL} = (176.2 – 6.6) = 169.6 ksi
    $f_{c(allow)}$ = 2.1 ksi (at transfer)

$$t_m = 6 + \frac{\pi(3)^2}{32.53} + \left[\left(\frac{169.6(346.6)}{107.5}\right) + 28.2\right]\frac{1}{2.1 \times 12}$$

$$t_m = 6 + .87 + 22.8 = 29.67 \text{ in}$$

Add 4″ for construction

$$t_O = 4 + 29.67 = 33.67 \text{ in (Use } 2' - 10'')$$

5.3 Other Requirements

Allowable compressive stresses may control the wall thickness under some load combinations other than mentioned above. See example as in Section 8.4.

6. TENDON MATERIAL STRESS – STRAIN CURVE

In order to assure that structural integrity is maintained, a realistic, reproducible stress–strain curve for the tendon material is required. The following method shows how to develop the stress–strain curve for ASTM A-421 wire. Similar results can be produced for other materials.

For ASTM A – 421 Wire

The stress–strain curve shown on Figure D-7 is developed from the minimum values given in ASTM A-421. It is assumed as follows:

$$E_1 = 29,000,000 \text{ psi}$$

1% elongation at 192 ksi (.8 fpu)

4% elongation at 240 ksi (fpu)

The transition curve between E_1 = 29,000 ksi and E_2 = 1600 ksi is a circle tangent to both lines, with one tangent at 1% elongation at 192 ksi. is assumed equal to 192 ksi.

Figure D-8 is an enlargement of the area of interest on Figure D-7. Three points are chosen from this curve. The two points A and B are then connected by straight lines. This is the assumed stress – strain curve for these calculations, and is shown in Figure D-9 in a more useful scale.

7. THERMAL STRESSES

To allow for initial reinforcing sizing, an estimate of the stresses in the steel due to thermal loads are required. These stresses may be obtained from the results of a finite element computer analysis for a similar structure previously analyzed. The analysis used should consider the effects of cracking of the concrete.

Assume the thermal rebar stress for accident conditions is between 20 ksi and 24 ksi.

8. SECTION EQUILIBRIUM AND STRAIN COMPATIBILITY

8.1 Discussion

To establish that structural integrity is maintained, it must be demonstrated that each section under each load combination will reach equilibrium without exceeding allowable material stresses and strains. Figure D-11 shows a concrete section, at equilibrium, subjected to prestress and tensile forces. ΔF represents the increase in prestress force due to the elongation of the concrete member caused by the applied tensile load, T. The prestress increase is called the pressurization effect.

This section of the design example shows methods which may be applied at various locations in the structure, to obtain material stresses and to show that section equilibrium is assured.

At sections where the load combination produces tension in the concrete, as in Sections 8.2 and 8.5 the graphical method shown includes the effects of pressurization. The stresses obtained are a good estimate of reality. In cases where the concrete remains in compression, as in Sections 8.3 and 8.6, the method of Figure D-11 is used to obtain the remaining compression in the concrete.

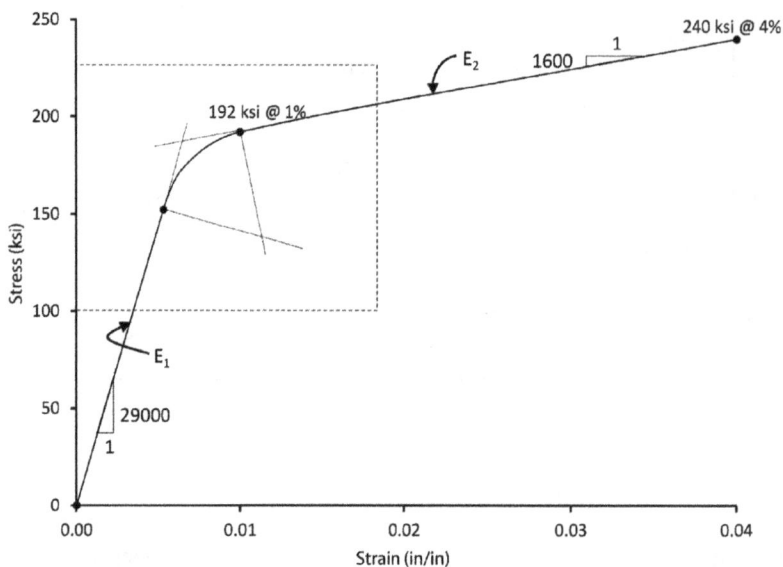

FIGURE D-7

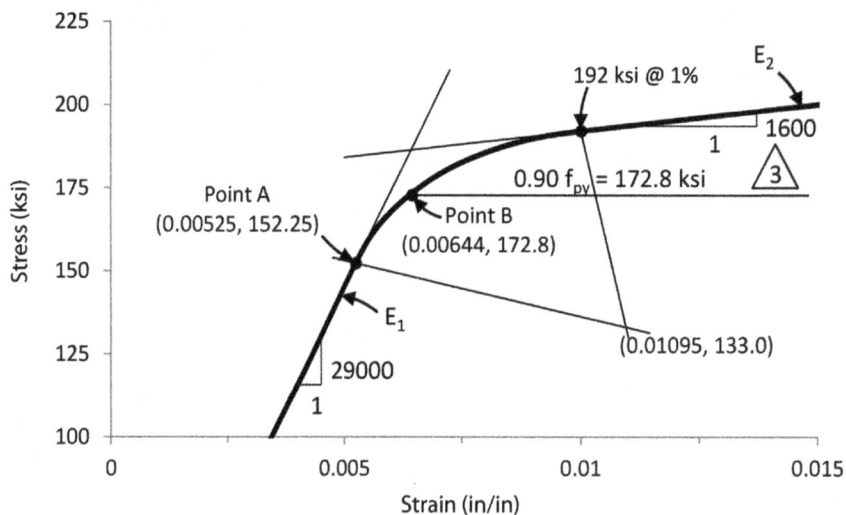

FIGURE D-8

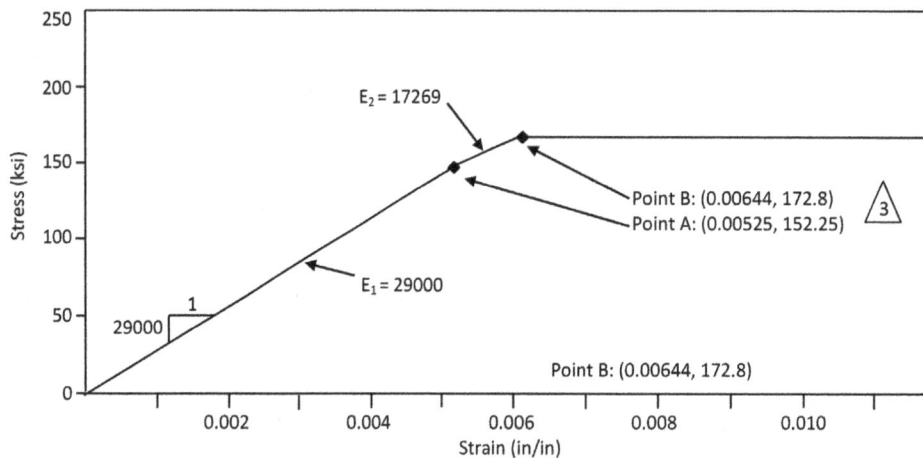

FIGURE D-9

In the following examples, the reinforcing steel is not included in the equilibrium calculation. This was done to show that equilibrium can be reached without the reinforcing so that mechanically spliced rebar is not required in the structure. It will be shown in the next section, Section 9, that, for this level of prestress, (1.2), mechanically spliced rebar are required in the hoop direction. In this case, the equilibrium calculation in the hoop direction could be redone to include the rebar to lower the section strain and provide a more efficient use of the steel.

8.2 Section Equilibrium Considering Pressurization

Σ external forces = Σ internal forces

$$F + \Delta F - T = C$$

$$F = A_c E_c \varepsilon_c$$

$$\Delta F = A_{ps} E_s \varepsilon_\Delta$$

$$C = A_c E_c (\varepsilon_c - \varepsilon_\Delta)$$

$$\therefore F + \Delta F - T = F - A_c E_c \varepsilon_\Delta$$

$$T = \varepsilon_\Delta (A_{ps} E_S + A_c E_c)$$

For $\varepsilon_\Delta \le \varepsilon_{Cc}$: $\varepsilon_\Delta = \dfrac{T}{A_{ps}E_S + A_{Cc}E_{cC}}$

so that $T \le F + AF$

where: F – force on section due to initial prestress
   ΔF – change in F due to straining of concrete under load T
   T – load on section from other loads
   C – force in concrete
   A_c – net concrete area
   A_{ps} – area of prestressing steel
   E_c – concrete modulus
   E_S – steel modulus
   ε_c – concrete strain from F (not including creep and shrinkage)
   ε_Δ – change in concrete strain due to T

8.3 Dome at Apex (Vertical Tendons)

Consider load combination: $1.0(D + F + L + T_a + R_a + 1.5P_a)$

Since the seismic force is zero at the apex of the dome, the above load combination generally produces the maximum stress in the tendon and reinforcing steel.

$$D = D_1 \sim 0, \ L = 0, \ R_a = 0$$

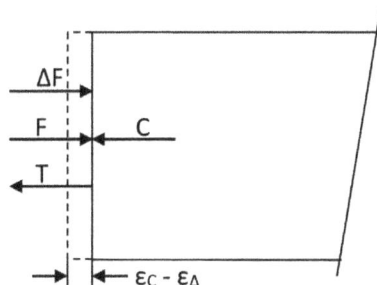

FIGURE D-10

Using minimum reinforcement $\rho = 0.25\%$

$$A_{Ss} = \frac{0.25}{100} \times 34 \times 12 = 1.02 \ \text{in}^2/\text{ft}$$

Use #10 @ 12" A_s = in²/ft
Initial Concrete Membrane Strain:

$$\varepsilon_{Cc} = \frac{F}{E_{Cc}A_{Cc}} = \frac{33,7000}{(4.1)(12)(34)} = 201 \ \text{micro strain}$$

Design information for reinforcing steel:
Thermal stress: f_t = 22.8 ksi
where f_t is the thermal stress in the reinforcing steel (refer to Section 7).
Reinforcing steel strain at 54,000 psi .9f_y:

$$\varepsilon_{.9fy} = \frac{(54 - 22.8)(1000)}{29} + 201 = 1276 \ \text{micro strain}$$

Reinforcing steel stress at zero micro strain in concrete:

$$f_s = 22.8 - .201(29) = 17.0 \ \text{ksi}$$

Initial tendon stress = 107.5 ksi
Investigation of tendon and reinforcing steel stress and strain:
Assume rebar does not resist membrane tension.

$$f_t = 22.8 \ \text{ksi}$$

$$U = 1.5P_a = 1.5 \frac{60 \times 65}{2 \times 1000}(144) = 421.2 \ \text{k/ft}$$

$$\frac{f_{ps}}{f_{se}}F + f_s A_s > 421.2$$

$$\frac{f_{ps}}{107.5}(337) + 0 = 421.2$$

$$f_{ps} = 421.2 \left(\frac{107.5}{337}\right) = 134.5 \ \text{ksi} < 152.25 \ \text{ksi (Point A of curve of Figure D-11)}$$

Section strain from Figure D-11 = 931 micro strain
Thermal rebar strain $-\dfrac{22.8}{29}(1000) = 786$ micro strain
Total rebar strain $= 786 + 931 = 1717 < \dfrac{1.5(60)(1000)}{29} = 3103$ micro strain
(Rebar stress limited to 1.5 times yield strain)

8.4 Vertical Tendon at Spring Line

Consider load combination $U = 1.0(D + F + L + T_a + R_a + 1.5P_a)$

Since seismic forces at this location are generally negligible, the above load combination generally governs.

$$D = D_2 = 28.2 \ \text{k/ft} \qquad L = 0 \qquad R = 0$$

Assume minimum reinforcement $\rho = 0.25\%$
$$A_s = \frac{0.25}{100} \times 44 \times 12 = 1.32 \ \text{in}^2$$

Use #11 @12" A_s = 1.56 in²/ft

Initial concrete membrane strain:

$$\varepsilon_c = \frac{F}{E_c A_c} = \frac{(141.7/107.5)(337000)}{4.1 \times 12 \times 40} = 206 \ \text{micro strain}$$

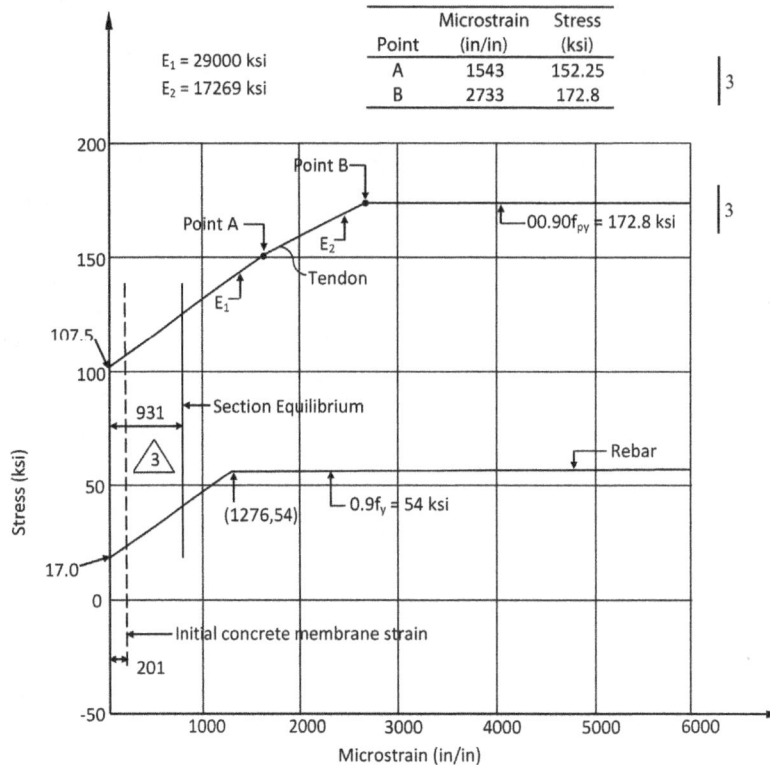

FIGURE D-11

Design information for reinforcing steel:

Thermal stress: $f_t = 23.4$ ksi

Reinforcing steel strain at 54,000 psi ($.9f_y$)

$$\varepsilon_{.9f_y} = \frac{(54-23.4)}{29}(1000)+206 = 1261 \text{ micro strain}$$

Reinforcing steel stress at zero micro strain in concrete:

$$f_s = 23.4 - .206(29) = 17.4 \text{ ksi}$$

Initial tendon stress = 141.7 ksi
Assume rebar does not resist membrane tension.

$$f_t = 23.4 \text{ ksi}$$

$$U = -28.2 + 1.5\frac{(60\times 65)(144)}{2(1000)} = 393 \text{ k/ft}$$

$$\frac{f_{ps}}{107.5}(337)+0 = 393$$

$$f_{ps} = 125.4 \text{ ksi} < 141.7 \text{ ksi}$$

Concrete is still in compression.

$$\varepsilon_\Delta = \frac{T}{(A_{ps}E_s + A_cE_c)} = \frac{393(1000)}{(3.14(29)+12(40)(4.1))}$$

$$= 190.9 < 206 \text{ micro stain}$$

where $A_{ps} = \frac{337}{107.5} = 3.14 \text{ in}^2/\text{ft}$

$F + \Delta F - T = C$
$F = \frac{141.7}{107.5}(337) =$ 444.17 k/ft

$\Delta F = 3.14(.1909)(29) =$ 18.13 k/ft
$T =$ 393.0 k/ft
$C =$ 69.30 k/ft (compression)

8.5 Vertical Tendon at Bottom of Wall

Check compression at bottom of wall.
Maximum tendon stress without losses:

$$f_{E.L.} = .7f_{pu} - 6.6 = 168.0 - 6.6 = 161.4 \text{ ksi}$$

where 6.6 ksi is the reduction in stress in the tendon due to elastic shortening.

$$F = \frac{161.4}{107.5}(337) = 506 \text{ k/ft (initial prestress force)}$$

$$D = D_3 = 116.7 \text{ k/ft}$$

$$E = N_{ve} = 146 \text{ k/ft (see Figure D-3)}$$

Consider load combination S = D+F+L+E

$$(D + F + E) = (116.7 + 506 + 146) = 768.7 \text{ k/ft}$$

$$f_c = \frac{768.7(1000)}{12\times 40} = 1601.5 \text{ psi} < .4f_c' = 2400 \text{ psi}$$

Note: On structures with higher seismic loads, this case may control the necessary wall thickness instead of the calculations in Section 5.

8.6 At the Face of the Buttress – Hoop-Direction

At anchor for tendon 1: $f_{se} = 168 - 34.5 = 133.5$ ksi
At anchor for tendon 2: $f_{se} = 139.8 - 34.5 = 105.3$ ksi
Consider load combination U = 1.0(D + F + L + T_a + R_a + 1.5P_a)

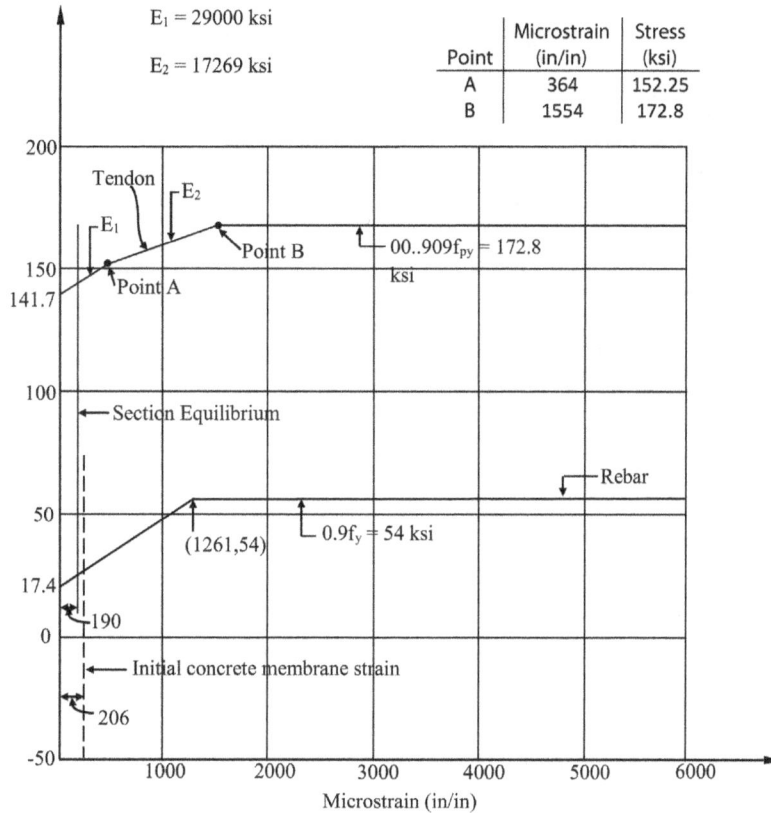

Point	Microstrain (in/in)	Stress (ksi)
A	364	152.25
B	1554	172.8

FIGURE D-12

$D = L = R_a = 0$ $\quad$ $A_s = 1.56\,in^2$ $\quad$ $\#11\ @\ 12''$

$$U = 1.5P_a = 1.5\frac{(60\times65)}{1000}(144) = 842.4\ k/ft$$

$$F = 1.5P_a = 1.2\frac{(60\times65)}{1000}(144) = 674\ k/ft$$

$$f_c = \frac{F}{A_c} = \frac{674}{12\times40}(1000) = 1404\ psi$$

$$\varepsilon_c = \frac{f_c}{E_c} = \frac{1404}{4.1} = 342\ micro\ strain$$

From Figure D-13, $f_{ps2} = 115.2$ ksi @ 342 micro strain

$$U = f_{ps1} + f_{ps2} + A_s f_s \geq 842.4$$

Trial No. 1 – Assume rebar does not resist membrane tension (see Figure D-13).

$$f_t = 23.4\ ksi$$

$$f_{ps1} = \frac{17.269}{1000}(\varepsilon'_{s1} - 647) + 152.25$$

$$f_{ps2} = \frac{29.00}{1000}(\varepsilon'_{s2} - 342) + 115.2$$

$$\frac{(674)}{133.5}\left[\frac{17.269}{1000}(\varepsilon_s - 647) + 152.25\right] +$$

$$\frac{(674)}{105.3}\left[\frac{29.00}{1000}(\varepsilon_s - 324) + 115.2\right] = 842.4 \times 2$$

$$0.0872\varepsilon_s - 56.41 + 769 + 0.1856\varepsilon_s - 60.14 + 737 = 842.4 \times 2$$

$$\varepsilon_s = \frac{295}{0.2728} = 1081\ micro\ strain$$

$$f_{ps1} = \frac{17.269}{1000}(1081 - 647) + 152.25 = 159.7\ ksi < 172.8\ ksi$$

$$f_{ps2} = \frac{29.000}{1000}(1081 - 342) + 115.2 = 136.63\ ksi < 152.25\ ksi$$

Rebars:
Section strain = 1081 micro strain
Thermal strain = $\frac{23.4}{29}(1000) = 807$ micro strain
Total strain = $(1081 + 807) = 1888$ micro strain < 3103 micro strain

8.7 Hoop Tendons at Bottom of Cylinder with Pressurization

Consider load combination $U = 1.0(D + F + L + T_a + 1.25P_a + 1.25E)$

$D = L = E = 0$ $\quad$ $A_S = 1.56\ in^2$ $\quad$ $\#11\ @\ 12''$

$$1.25P_a = 1.25\frac{(60\times65)}{1000}(144) = 704\ k/ft$$

$$F = 1.2P_a = \frac{702(1.2)}{1.25} = 674\ k/ft$$

$$\varepsilon_c = \frac{F}{A_c E_c} = \frac{674}{4.1(12)(40)} = 342\ micro\ strain$$

$$A_{ps} = \frac{2(674)}{133.5 + 105.3} = 5.64\ in^2/ft$$

$E_1 = 29000$ ksi

$E_2 = 17269$ ksi

(342,115.24

Point	Microstrain (in/in)	Stress (ksi)
A_1	647	152.25
B_1	1837	172.80
A_2	1612	152.25
B_2	2802	172.80

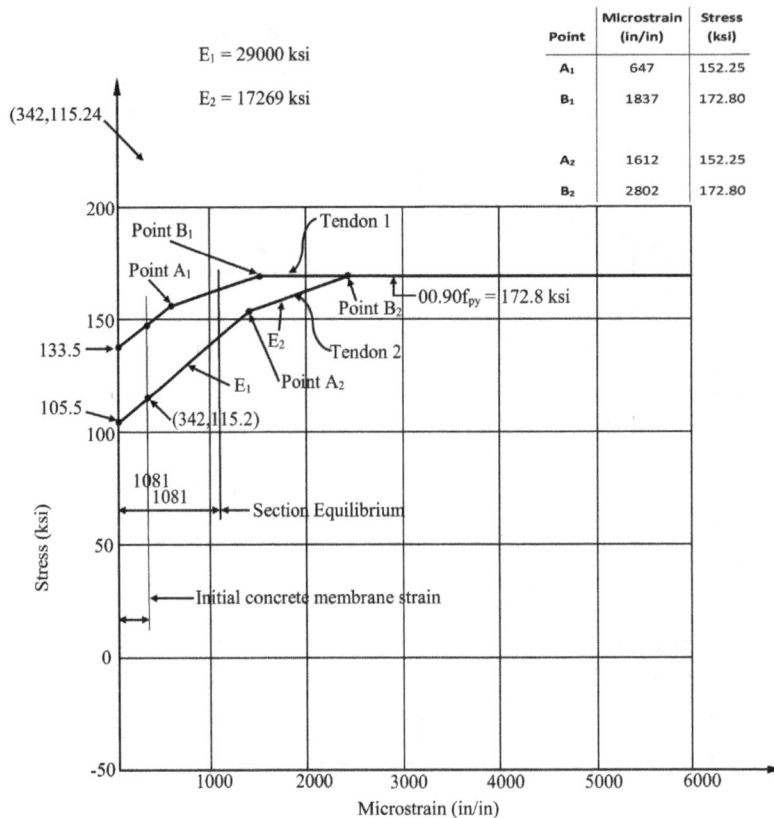

FIGURE D-13

$$\varepsilon_\Delta = \frac{T}{E_s A_{ps} + E_c A_c} = \frac{702(1000)}{29(5.64) + 4.1(12)(40)}$$

$$= 329 \text{ micro strain}$$

$$\varepsilon_\Delta = 329 < \varepsilon_c = 342 \text{ micro strain}$$

Concrete is still in compression

$$F + \Delta F - T = C$$

$$674 + A_{ps}E_s\varepsilon_\Delta - 702 = C$$

$$674 + \frac{5.64(29)(329)}{1000} - 702 = C = 25.8 \text{ k/ft}$$

$$f_{ps1} = 133.5 + \frac{329(29)}{1000} = 143.0 \text{ ksi} \quad f_s = 23.4 \text{ ksi}$$

$$f_{ps2} = 105.3 + \frac{329(29)}{1000} = 114.8 \text{ ksi}$$

9. Tangential Shear Reinforcement

This article shows the method for sizing reinforcing steel to resist seismic tangential shear loads.

From Appendix C of this topical report:

$$A_h = \frac{N_h + (V_u^2 + N_{he}^2)^{1/2}}{0.9 f_y} \tag{9-b}$$

Consider load combination:

$$U = 1.0(D + F + L + T_a + 1.25P_a + 1.25E_o)$$

Continuing with the example in Section 8.7

$$N_h = -25.8 \text{ k/ft (including pressurization effect)}$$

$$V_u = 1.25 \times 69.92 = 87.4 \text{ k/ft (see Table D-2)}$$

$$N_{he} = 0$$

$$A_h = \frac{-25.8 + 87.4}{(.9 \times 60) - 23.4} = 2.01 \text{ in}^2/\text{ft}$$

Note: 23.4 ksi is the thermal stress in rebar; use #14 @ 13 in. in hoop direction.

Mechanical splices are required for these rebar because they are utilized to carry the tangential shear load.

Also note, that rebar required to carry secondary moments at the bottom of the cylinder should be in addition to the above.

If the prestress level in the hoop direction is increased to such a value that A_h becomes zero in the above example, then the thermal load mainly governs the rebar requirement. Under this condition, rebar are not required to be mechanically spliced.

[D-2] REINFORCED CONCRETE CONTAINMENT SHELL DESIGN FOR PRIMARY LOADS AND THERMAL EFFECTS BY THEODORE E. JOHNSON, TJBG CONSULTING INC.

REINFORCED CONCRETE CONTAINMENT SHELL DESIGN FOR PRIMARY LOADS AND THERMAL EFFECTS

Prepared by: Theodore Johnson
TJBG Consulting Inc.
Reviewed by: Gunnar Harstead
Harstead Consulting

Illustrated below is what can be considered as an initial proportioning of the basic Reinforced Concrete Containment [RCC] Shell. Only the design conditions that typically control the design are considered in what is presented below. The Containment Designer is required to address all the various conditions contained in the code. The material presented here only considers membrane type loads. The effects of membrane and bending loads at areas such as discontinuities are not considered in this example nor is radial shear.

TABLE OF CONTENTS

6.1.1 Required Reinforcing for Tangential Shear and Membrane Forces is repeated below for ease of reference:

CC-3521.1.1 Reinforced Concrete

(a) Required area of orthogonal (hoop and meridional) reinforcement, with or without inclined reinforcement, provided for combined tangential shear and membrane strength shall be computed by:

$$A_{sh} + A_{si} = [N_h + (N_{hl}^2 + V_u^2)^{1/2}] / 0.9* f_y \qquad (12)$$

$$A_{sm} + A_{si} = [N_m + (N_{ml}^2 + V_u^2)^{1/2}] / 0.9* f_y \qquad (13)$$

(b) Any combination of orthogonal and inclined reinforcement as required for strength according to equations (12) and (13), and as required to control shear deformations may be used with the following limits on maximum shear force.

(1) Tangential shear strength provided by orthogonal reinforcement V_{so} and computed by

$$V_{so} = V_u - 0.9*f_y*A_{si} \qquad (14)$$

shall not be greater than $0.2* f'_c *bt$.

(2) Tangential shear force V_u shall not exceed $0.4* f'_c *bt - V_{so}$ where V_{so} is computed according to Equation (14) where

A_{sh} = area of bonded reinforcement in the hoop direction (in.^{2}/ft)

A_{si} = area of bonded reinforcement in one direction of inclined bars at 45 deg to horizontal [in.^{2}/ft (mm^2 / m) along a line perpendicular to the direction of the bars]. Inclined reinforcement shall be provided in both directions (in.^{2}/ft)

A_{sm} = area of bonded reinforcement in the meridional direction (in.^{2}/ft)

N_h and N_m = membrane force in the hoop and meridional direction, respectively, due to pressure and dead load. N_h and N_m are positive when in tension and negative when in compression.

N_{hl} and N_{ml} = membrane force in the hoop and meridional direction, respectively, from lateral load such as earthquake, wind, or tornado loading. When considering earthquake loading, this force is based on the square root of the sum of the squares of the components of the two horizontal and vertical earthquakes. The force is always considered as positive.

V_{si} = tangential shear strength provided by inclined reinforcement

V_{so} = tangential shear strength provided by orthogonal reinforcement

V_u = the peak membrane tangential shear force resulting from lateral load such as earthquake, wind, or tornado loading. When considering earthquake loading, this force is based on the square root of the sum of the squares of the components of the two horizontal and vertical earthquakes. The shear force shall be considered as positive. All forces and strengths are expressed in kips/ft. The strain compatibility of the concrete and reinforcement shall be checked to ensure that the strain allowables in CC-3422 are not exceeded for orthogonal and inclined reinforcement.

Listed below is an excerpt from NUREG-0800, Rev2-[Month] 2007 Pg 3.8.1-18

The tangential shear strength provided by orthogonal reinforcement should be limited to the following:

$[0.833*(f'_c)^{1/2} (MP_a); \quad [10*(f'_c)^{1/2}](psi)$

where the value of f'_c is in units of MPa and psi in the first and second expression, respectively, in accordance with the ASME Code.

In this design example the SSE Cylinder Tangential Shear at Base = 260 k/ft. The orthogonal reinforcement is allowed to carry:

$$V_{so} = 0.2*f'_c*bt = (0.2*4000*12*48)1000 = 461 \text{ k/ft}$$

Now using the NUREG value:

$$V_{so} = [10*(f'_c)^{1/2}]*bt = ([10*(4000)^{1/2}]*12*48)/1000 = 364 \text{ k/ft}$$

It would appear that a containment would have to be located in a fairly high seismic region before inclined reinforcing would be required.

7. Reinforcing Determination Primary Loads
   7.1 Factored Load Category
       7.1.1 Abnormal: $1.0*D + 1.5*P_a$
   At Mid Height Hoop

$A_s = (1.0*D + 1.5*P_a)/(0.9*f_y) = (0.0 + 1.5*562)/(0.9*60) = \underline{15.61 \text{ in}^2/\text{ft}}$

At Dome Apex

$A_s = (1.0*D + 1.5*P_a)/(0.9*f_y) = (0.0 + 1.5*281)/(0.9*60) = \underline{7.81 \text{ in}^2/\text{ft}}$

At Cylinder Meridional at Base

$A_s = (1.0*D + 1.5*P_a)/(0.9*f_y) = (-122 + 1.5*281)/(0.9*60) = \underline{5.45 \text{ in}^2/\text{ft}}$

7.1.2 Abnormal/Severe Environmental: $1.0*D + 1.25*P_a + 1.25*E_o$
At Base Hoop

$N_h = 1.25*562 = 703 \text{ k/ft}$

$(N_{hl}^2 + V_u^2)^{1/2} = 1.25*120 = 150 \text{ k/ft}$

$A_{sh} = [N_h + (N_{hl}^2 + V_u^2)^{1/2}]/0.9*f_y$

$\qquad = [703 + 150]/(0.9*60) = 15.8 \text{ in}^2/\text{ft}$

At Cylinder Meridional at Base

$N_m = -122 + 1.25*281 = 229 \text{ k/ft}$

$(N_{ml}^2 + V_u^2)^{1/2} = ((1.25*240)^2 + (1.25*120)^2)^{1/2}$

$\qquad = 335 \text{ k/ft}$

$A_{sm} = [N_m + (N_{ml}^2 + V_u^2)^{1/2}]/0.9*f_y$

$\qquad = [229 + 335]/(0.9*60) = 10.44 \text{ in}^2/\text{ft}$

7.1.3 Abnormal/Extreme Environmental: $1.0*D + 1.0*P_a + 1.0*E_{ss}$

At Base Hoop

$N_h = 1.0*562 = \text{k/ft}$

$(N_{hl}^2 + V_u^2)^{1/2} = 1.0*260 = 260 \text{ k/ft}$

$A_{sh} = [N_h + (N_{hl}^2 + V_u^2)^{1/2}]/0.9*f_y$

$\qquad = [562 + 260]/(0.9*60 = 15.22 \text{ in}^2/\text{ft})$

At Cylinder Meridional at Base

$N_m = -122 + 1.0*281 = 159 \text{ k/ft}$

$(N_{ml}^2 + V_u^2)^{1/2} = ((1.0*525)^2 + (1.0*260)^2)^{1/2} = 586 \text{ k/ft}$

$A_{sm} = [N_m + (N_{ml}^2 + V_u^2)^{1/2}]/0.9*f_y$

$\qquad = [159 + 586]/(0.9*60) = 13.8 \text{ in}^2/\text{ft}$

7.2 Service Loads [Test Condition]
    7.2.1 Structural Integrity Test: $1.0*D + 1.15*P_a$
    At Mid Height Hoop

$A_s = (1.0*D + 1.15*P_a)/(0.75*f_y) = (0.0 + 1.15*562)/(0.75*60) = \underline{14.36 \text{ in}^2/\text{ft}}$

At Dome Apex

$A_s = (1.0*D + 1.15*P_a)/(0.75*f_y) = (0.0 + 1.15*281)/(0.75*60) = \underline{7.18 \text{ in}^2/\text{ft}}$

At Cylinder Meridional at Base

$A_s = (1.0*D + 1.15*P_a)/(0.75*f_y) = (-122 + 1.15*281)/(0.75*60) = \underline{4.47 \text{ in}^2/\text{ft}}$

7.3 Reinforcing Summary
    (a) Apex: 7.81 in²/ft
    (b) Hoop Mid-height: 15.61 in²/ft
    (c) Hoop Near Base: 15.8 in²/ft
    (d) Cylinder Meridional at Base: 13.8 in²/ft

8. Primary Loads Combined With Thermal Effects
   The following example is typical of a reinforced containment when subjected to accident pressure and a thermal gradient. For illustration, the Poisson effect will be neglected since it is only applicable to the liner plate, but not to the reinforcing steel. A very high liner temperature will be used to illustrate yielding and the hoop direction will be considered at mid-height.

8.1 Basic Information:
    (a) Wall thickness 48 in
    (b) Reinforcing Area 15.61/2 = 7.81 in²/ft each face, A_s & A'_s, assumed yield = 54 ksi
    (c) Liner Plate Thickness 3/8 in & Area = 0.375*12 = 4.5 in²/ft, A_L
    (d) Primary Load 1.5 P_a = 843 k/ft
    (e) Temperatures
        * Liner 400 degrees F
        * Inside Concrete 100 degrees F
        * Outside Concrete 0 degrees F
    (f) a [Coef of Expansion] = 6.0×10^{-6}, $E_s = 30 \times 10^6$, $a*E_s = 180$

8.2 Analysis
    The technique will setup the equilibrium equation and then iterate until the final solution is found.
    * The Figure below shows the initial Stress Diagram

Stress Diagram

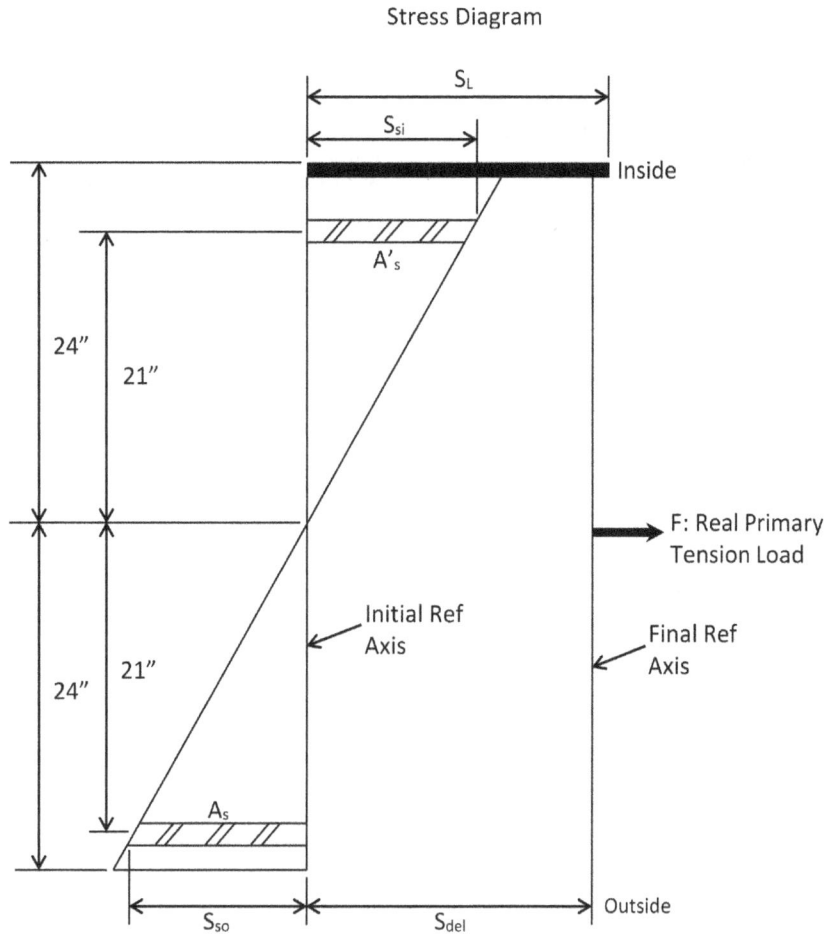

Stresses from the thermal gradient prior to concrete cracking and relative to the reference axis are as follows:

Outside Reinforcing Stress [tension]

$S_{so} = (21/24)*(100/2)*180 = 7875$ psi

Inside Reinforcing Stress [compression]

$S_{si} = -(21/24)*(100/2)*180 = -7,875$ psi

Liner Plate Stress [compression]

$S_L = -(400 - (100/2))*180 = -63,000$ psi

S_{del} = Reference Axis Stress Shift

It was previously shown that with the load of 843 k/ft and both inside and outside reinforcing at 7.81 in²/ft that under the primary load the design was adequate and the stresses were at the assumed yield of 54,000 psi.

With this large tension load, the concrete will crack completely through the thickness and need not be considered.

GENERAL FORCE EQUILIBRIUM EQUATION WITH NO YIELDING AND CONCRETE FULLY CRACKED

Outside Reinforcing	Inside Reinforcing	Liner Plate	Real Force

$$(S_{so} + S_{del})*A_s - (S_{si} - S_{del})*A'_s - (S_L - S_{del})*A_L = F$$

Find the tension force that causes yielding in the outside reinforcing, i.e.,

$$S_{del} = 54,000 - 7875 = 46,125 \text{ psi}$$

$$(7875 + 46,125)*7.81 - (7875 - 46,125)*7.81 - (63,000 - 46,125)*4.5 = F'$$

$$F' = 644,535 \text{ lbs}$$

State of Stress and Force for F' = 644535 lbs

	Stress [psi]	Force [kips]
Liner Plate	−16,875	−75.9
Inside Reinforcing	38,250	298.7
Outside Reinforcing	54,000	421.7

Find the tension force that causes yielding in the inside reinforcing, i.e.,

$$S'_{del} = 54,000 - 38,250 = 15,750 \text{ psii}$$

$$(54000)*7.81 - (7875 - 46,125 - 15,750)*7.81 - (63,000 - 46,125 - 15,750)*4.5 = F''$$

$F'' = 838,418$ lbs

	Stress [psi]	Force [kips]
Liner Plate	−1125	−5.1
Inside Reinforcing	54,000	421.7
Outside Reinforcing	54,000	421.7

At this point, the liner will relieve the −1125 psi since all the reinforcing is at yield, there is no resistance to the liner expansion, and the final condition is:

	Stress [psi]	Force [kips]
Liner Plate	0	0.0
Inside Reinforcing	54,000	421.7

| Outside Reinforcing | 54,000 | 421.7 |

The final Strains in the Liner and the Reinforcing are as follows:

	Strain (in/in)
Liner Plate	0
Inside Reinforcing	$[54,000 + 1125]/(30 \times 10^6) = 1836 \times 10^{-6}$
Outside Reinforcing	$[54,000 + 15,750 + 1125]/(30 \times 10^6) = 2363 \times 10^{-6}$

In this example yield has been assumed to correspond to a stress level of 54,000 psi, and the yield strain would be:

$[54000]/(30 \times 10^6) = 1800 \times 10^{-6}$. Therefore the outside reinforcing is at $(2,363/1800) = 1.31$ times yield.

APPENDIX E—TANGENTIAL SHEAR CONSIDERATIONS

[E-1] SUMMARY CONCRETE CONTAINMENT VESSEL DESIGN FOR TANGENTIAL SHEAR LOADS, T.E. JOHNSON AND D.F. GREEN, BECHTEL POWER CORPORATION, SMIRT 4, 1977 [J3/4].

SUMMARY

This paper presents the historical development of tangential shear considerations for concrete containments in the United States.

About forty years ago, the first major generation of concrete containments were under design in the United States. In general, the reinforced containments used both an orthogonal (hoop, vertical) and inclined (diagonal) reinforcing system. The orthogonal reinforcing was designed to resist membrane loads and the inclined system resisted the tangential shear. Prestressed containments allowed a concrete principal tension stress for various loading combinations which included earthquake loads.

When the first ASME Section III, Division 2 Containment Code was issued for use it had provisions for considering tangential shear in reinforced containments, but no provisions were given for prestressed containments. The provisions for reinforced containments were extremely conservative. Another requirement for reinforced containments requires that strain compatibility be considered and no yielding was permitted in the inclined reinforcing.

Later, there were several code changes that were approved for reinforced containments and provisions were being added for prestressed containments and the bases for these changes are covered in this paper.

[E-2] REPORT TO ASME-ACI 359 SUBGROUP ON DESIGN, TANGENTIAL SHEAR CODE PROVISIONS BY TASK GROUP ON SHEAR, R. G. OESTERLE J. A. CURTIN, T. E. JOHNSON, P. SHUNMUGAVEL, A. WALSER, R. N. WHITE, NOVEMBER 7, 1986

TABLE OF CONTENTS

1. INTRODUCTION

1.1 Background

Concrete structures in nuclear power plants have been used extensively since the beginning of the nuclear power industry. Initially, concrete was used for radiation shielding. However, use of reinforced and prestressed concrete structures as pressure containments was started in the 1960's. Although concrete had been used in safety-related structures for many years, use of concrete in pressure vessels was a new concept. The size, shape, and possible stress states in containments produced many unique problems for both design and construction. Because of the importance of containments, a high degree of reliability was sought in solving these problems. This philosophy has sometimes led to cumbersome designs.

A primary example of difficulty is design and construction problems produced by the questioned capacity of concrete to transfer shear stress while in a state of biaxial tension. Concrete containments in the United States are designed to resist a combination of biaxial tension caused by internal pressure, and tangential shear caused by earthquakes. To resist internal pressure, reinforcement is generally placed in an orthogonal pattern of vertical and horizontal bars. Use of the same orthogonal reinforcement to resist earthquake forces requires that tangential shear stresses be transferred across open orthogonal cracks. The shear transfer capacity across the open cracks requires experimental verification.

1.2 Review of Experimental Work

The capacity for force transfer across an open crack in reinforced concrete has been the subject of a number of experimental investigations conducted during the past three decades. These investigations can be categorized by the type of specimen used. Specimens have included cracked joints in pavements, predefined cracks in unreinforced concrete, predefined cracks in concrete with internal reinforcement crossing the cracks, and randomly induced cracks in reinforced concrete panels.

Some of the early testing programs were conducted to evaluate the effectiveness of shear transfer by aggregate interlock across open control joints in concrete pavement. Experimental work by Colley and Humphrey[1] included alternating loads on each side of an open joint which simulated a wheel load crossing the joint. Parameters studied were aggregate size and joint opening width. Results indicated that joints with opening widths up to 0.065 in. work initially 80% as effectively as a closed joint. Effectiveness was evaluated by comparison of joint deflections. With cyclic loading, joint effectiveness decreased as a function of increase in joint opening width. As an example, the effectiveness of a joint with a width of 0.035 in. was reduced to approximately 50% of that of a closed joint after 500,000 cycles. It should be emphasized, however, that effectiveness was judged by comparison of <u>deflections</u> across the joint. Therefore, although the results of this study indicated that cyclic load reduced the joint stiffness, results did not necessarily show strength experienced a similar decrease.

Several investigators[2,3,4] have tested unreinforced concrete specimens with predefined cracks restrained in very stiff test frames. The objective was to evaluate aggregate interlock across cracks with <u>constant</u> width. Results of testing by Paulay and Loeber[3] indicated that although the stiffness was definitely decreased as crack widths were increased from 0.005 in. to 0.020 in., the maximum shear stress transferred across the 0.020 in. wide crack was only slightly less than the maximum stress across crack widths of 0.005 in. and 0.010 in. Also, the maximum stress in all specimens was greater than 1000 psi. This is a very high stress for interface shear transfer and not likely to be encountered in real structures. A stress of 1000 psi is higher than allowable by the ACI Building Code.[5] Also, by maintaining a constant width, the crack was essentially provided with an unrealistic "infinite stiffness" for deformation normal to the crack.

To more realistically model the stiffness normal to cracks, specimens with a predefined crack in unreinforced concrete restrained by external rods were used in a number of experimental programs.[6–8] Initial crack widths up to 0.030 in. were used. These tests demonstrated that aggregate interlock is an effective means of transferring shear stress across cracked concrete surfaces. The testing by White and Holley[6] is the basis for the current ASME-ACI

Code provision[9] allowing a conservative nominal tangential shear stress of up to 160 psi to be resisted by orthogonal reinforcement across open cracks in containments. However, test specimens with external rods did not accurately model the coupled effects between aggregate interlock and restraint from reinforcement embedded in the concrete across the crack.

Other researchers[10,11,12,13] have used specimens with embedded reinforcement crossing predefined cracks. Testing programs included specimens subjected to reversing load.

Mattock[12] determined that reversing load decreases the strength of the interface shear transfer mechanism to approximately 80% of the monotonic strength. Also, increase in initial crack width decreased shear transfer strength. A specimen with an initial crack width of 0.025 in. showed a strength reduction of approximately 15% compared with a specimen with an initial crack width of 0.015 in. However, the strength of the specimen with an initial crack width of 0.025 in. still exceeded ACI Building Code[5] allowable shear friction strength of 1.4 ρf_y. Also, the strength of this specimen was 660 psi. This is a very high stress for interface shear transfer and not likely to be encountered in real structures.

Test programs in References 2 through 13 were conducted to evaluate shear transfer in the plane of one predefined crack with an initial opening or with tension applied perpendicular to the crack. Although these tests contributed greatly to knowledge of the detailed behavior of interface shear transfer, the test specimens modeled a relatively artificial situation. Containment walls contain orthogonally cracked elements resulting from membrane tension caused by pressurization. The capacity of shear transfer mechanism in the concrete under a state of biaxial tension had not yet been verified experimentally. Therefore, current ASME ACI Code provisions[9] still require all but a nominal amount (up to 160 psi) of tangential shear to be resisted by inclined reinforcement. The inclined reinforcement is difficult to fabricate. It also adds significant congestion and inhibits concrete placement.

To provide experimental verification of the behavior of concrete containment walls subjected to biaxial tension and tangential shear forces, test programs have been conducted by the Construction Technology Laboratories (CTL) of the Portland Cement Association,[14,15] Cornell University,[16,17] and the University of Toronto.[18] Specimens were concrete panels containing internal reinforcement in two or four directions. Cracking was induced at random locations by tensioning the elements. A membrane shear stress in orthogonal directions was simulated in these specimens rather than applying a direct shear stress across one localized plane.

Results from these concrete panel tests and other testing and analysis, conducted primarily in Japan,[19-27] indicate that the current ASME-ACI[9] Code provisions for tangential shear strength are very conservative. Significantly higher shear stresses can be allowed without inclined reinforcement. The purpose of this report is to present recommended design criteria for tangential shear based on available test data.

2. NOMENCLATURE

The common approach to design for shear in reinforced concrete is to allocate some strength to the "concrete contribution," V_c. The remaining required strength is provided by reinforcement, V_s. The "concrete contribution" consists of shear through a compression zone, aggregate interlock, and dowel action. Although the steel used for V_s has some indirect influence on V_c, because of dowel action, there is no reinforcement directly provided for V_c.

Nomenclature used in the current ASME ACI code[9] is inconsistent with this approach and also inconsistent within itself. For tangential shear, V_c is defined in the current code as shear force carried by concrete. However, V_c is calculated as the strength provided by orthogonal (meridional and hoop) reinforcement. The "steel contribution," V_s, is provided by inclined reinforcement and no "concrete contribution" is considered. However, for radial and peripheral shear, V_c is a "concrete contribution" in that no reinforcement is required for this portion of shear strength.

Because of inconsistency in terminology, there is some confusion as to what V_c means. It is recommended that nomenclature be redefined to be consistent with other codes and within the ASME-ACI Code.[9] Toward this goal, it was recommended by the Task Group on Shear that the following definitions and relationships be used.

v_c = Tangential shear strength provided by concrete
v_{so} = Tangential shear strength provided by orthogonal (meridional and hoop) reinforcement
v_{si} = Tangential shear strength provided by inclined reinforcement
$v_u = v_s + v_c$
$v_s = v_{so} + v_{si}$

These changes were included in Subgroup on Design Action Item D83-1, Joint Committee Item JC 83-16. This item was passed by the ASME B&PV Committee in November 1983 and included in the summer 1984 Addenda to the Code.

3. STRENGTH PROVISIONS

As stated under Review of Experimental Work in this report, it is not likely that a localized plane subjected to only shear stress in one direction and of the magnitude measured in some of the shear test specimens (600 to 1000 psi) would be encountered in a real structure.

Results of panel tests[14–17] demonstrated that the interface shear transfer across open cracks is adequate to resist loads up to a level of stress where diagonal cracking occurs.

After diagonal cracking, a truss[15] mode of shear transfer takes over and orthogonal cracks are closed by the resulting diagonal compression. Testing of reinforced concrete panels indicates the interface shear transfer strength is adequate in specimens subjected to a biaxial tension steel stress up to 90% of specified yield. These specimens were subjected to cyclic shear with initial orthogonal crack widths up to 0.040 in. After diagonal cracking, shear strength is limited by either yield of the reinforcement in a diagonal tension mode or crushing of the concrete from diagonal compression.

3.1 Diagonal Tension

All specimens tested by CTL and Cornell[14–17] lost load capacity by yielding of reinforcement across a diagonal crack. Figure E-1 shows potential yield planes across the specimens tested by CTL. Using the free-body diagram shown in Figure E-2 the following equilibrium equations for yielding of reinforcement in the weaker of the horizontal or vertical directions are derived.

$$A_s f_y = N + V_{max} \qquad (1)$$

$$\frac{A_S}{bt} f_y = \frac{N/A_S}{bt/A_S} + \frac{V_{max}}{bt} \qquad (1a)$$

$$v_{max} = \frac{V_{max}}{bt}, \quad \rho = \frac{A_S}{bt}, \quad f_s = \frac{N}{A_S}$$

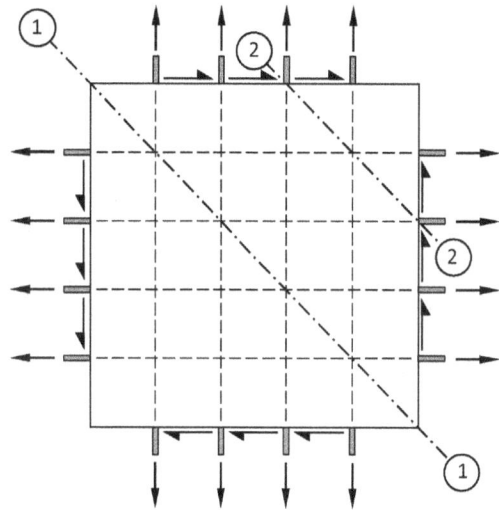

FIGURE E-1   PLANES THROUGH SPECIMEN FOR DIAGONAL TENSION STRENGTH.

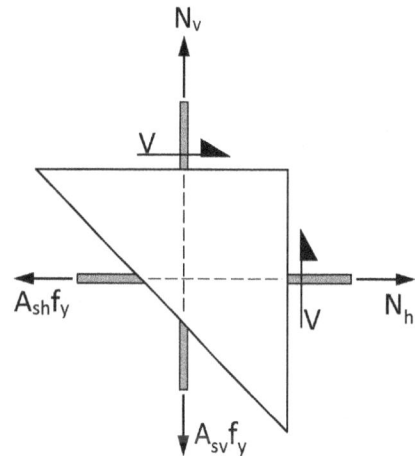

FIGURE E-2   FREE-BODY DIAGRAM FOR DIAGONAL TENSION EQUILIBRIUM

$$\rho f_y = \rho f_s + V_{max} \qquad (2)$$

$$v_{max} = \rho f_y \left(1 - \frac{f_s}{f_y}\right) \qquad (2a)$$

The design equations for tangential shear in the current ASME-ACI Code[9] are based on equilibrium and are expressed in a form similar to Equation (1) with a strength reduction factor of 0.9 for reinforcement yield stress. Figure E-3 shows maximum observed shear stress, v_{max}, for specimens tested by CTL and Cornell, versus calculated effective diagonal tension strength. The dashed line represents the simple diagonal tension equilibrium equation in the form of Equation (2a) with the strength reduction factor of 0.9. Figure E-3 shows that there is significant shear strength under biaxial tension. No specimens failed due to sliding shear or dowel splitting. Shear transfer across the orthogonal cracks was adequate for the shear stresses sustained by the specimens up to a diagonal tension failure.

The equilibrium equation with a reduction factor of 0.9 encompasses all but one data point. The first reversing load specimen

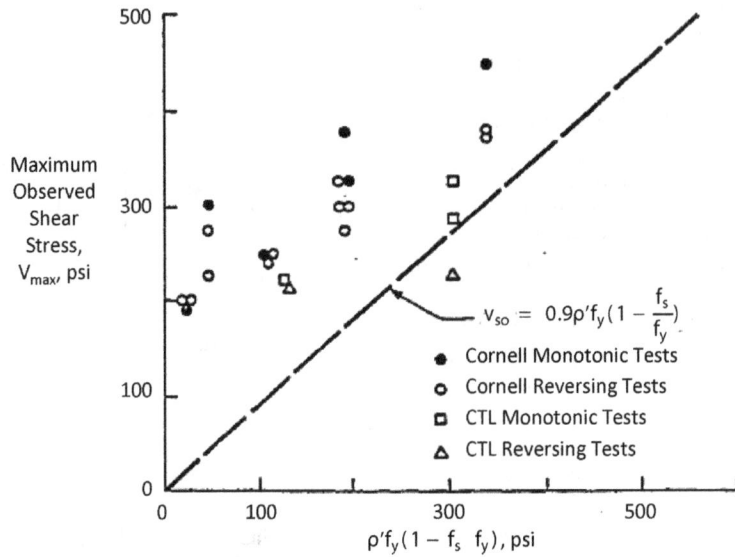

FIGURE E-3   DIAGONAL TENSION STRENGTH.

tested in the large-scale program lost shear capacity at a load lower than that predicted by simple equilibrium. This failure was attributed to stress concentrations in the loading system and should not be taken as indicative of the specimen diagonal tension strength.

Figure E-4 indicates a summary of Japanese test results.[28] The diagonal line represents simple diagonal tension equilibrium. As shown in this figure, the equilibrium equation is confirmed by Japanese testing up to a shear stress level of approximately 20. The upper limit for shear stress is discussed under Maximum Strength in this report.

It is the recommendation of the Task Group on Shear that the equilibrium equations similar to those currently in the code[9] continue to be used for design of tangential shear reinforcement with the following exception.

1. The term V_{so} replaces the term V_c as discussed in the section on Nomenclature in this report. (Note: this change is included in the Summer 1984 Addenda as previously discussed)

2. The normal and shear forces resulting from earthquake loading be combined with a Square Root of the Sum of Squares (SRSS) approach similar to that in the current code case.[29] This SRSS approach is based on calculations for maximum combination of N and V that can occur anywhere along the circumference of the containment as described in Appendix A of this report.

3. Required area of orthogonal (hoop and meridional) reinforcement, with or without inclined reinforcement, provided for combined membrane and tangential shear strength shall be computed by:

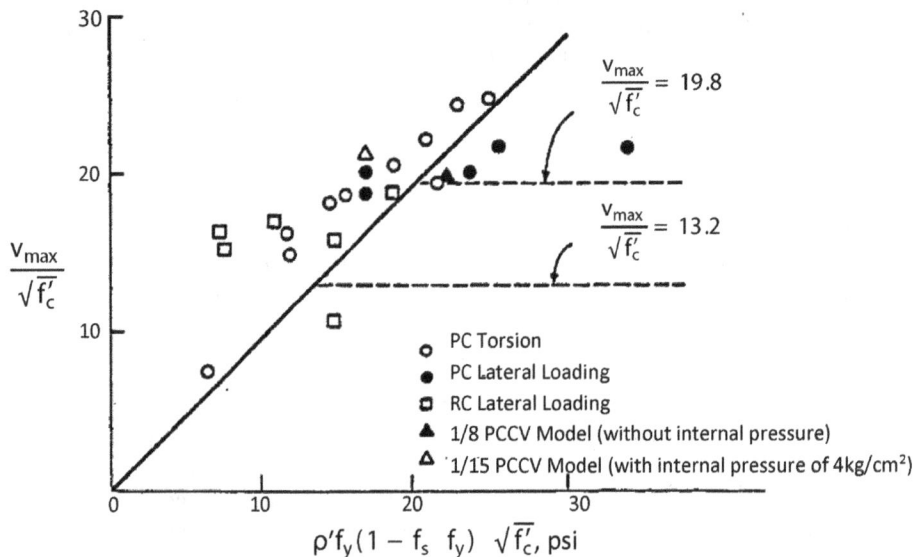

FIGURE E-4   SHEAR STRENGTH OF SPECIMENS TESTED IN JAPAN.[28]

$$A_{sh} + A_{si} = \frac{N_h + [N_{h1}^2 + V_u^2]^{1/2}}{0.9f_y} \qquad (3)$$

$$A_{sm} + A_{si} = \frac{N_m + [N_{m1}^2 + v_u^2]^{1/2}}{0.9f_y} \qquad (4)$$

4. Any combination of orthogonal and inclined reinforcement as required for strength according to Equations (3) and (4), and as required to control shear deformations, may be used. However, limits must be placed on maximum shear strength provided by the orthogonal reinforcement V_{so}, and maximum total shear V_u, so that reinforcement will yield before crushing of the concrete in compression can take place. These limits are discussed under maximum strength in this report.

3.2 Concrete Contribution

3.2.1 Reinforced Concrete

Currently, there is no "concrete contribution" allowed when designing for tangential shear in reinforced concrete containments under combined membrane tension and shear. The V_c in the current ASME-ACI Code is actually a V_{so} as discussed under Nomenclature in this report. Also, it is noted that recommended Equations (3) and (4) do not include any "concrete contribution" term.

The difference between observed strength and calculated diagonal tension strength shown in Figures E-3 and E-4 represent additional strength due to a "concrete contribution" and strain hardening of reinforcement. Figure E-3 suggests that at low levels of $\rho' f_y (1-f_s/f_y)$ (high levels of biaxial tension), there is significant "concrete contribution," V_c. However, it is apparently reduced by reversing loads and by increasing $\rho' f_y (1-f_s/f_y)$ (decreasing biaxial tension). The apparent loss of V_c with decreasing biaxial tension is probably due to the influence of boundary conditions and methods of loading the test specimens. As biaxial tension is decreased, the diagonal tension shear strength increases. Therefore, the level of shear stresses

carried by the specimens increases. The boundary conditions and loading methods would have a larger influence on measured strength at the higher levels of shear stress indicating an unrealistic loss of V_c. However, to stay within the limits of experimental data, it is conservatively recommended at this time that $V_c = 0$ for load cases that include membrane tension in reinforced concrete containments.

3.2.2 Prestressed Concrete

In reinforced concrete containments, orthogonal cracks generally occur during the structural integrity test. Therefore, reinforced concrete elements will behave as cracked sections for any further loading. Since V_c has traditionally been associated with the shear force causing diagonal cracking, the fact that a containment is pre-cracked has always been a reason for questioning the "concrete contribution" V_c in reinforced concrete containments. As stated above, although there is some experimental evidence that a significant V_c exists, it is conservatively recommended that $V_c = 0$ for reinforced concrete containment.

A prestressed containment, however, should not crack significantly during the structural integrity test. The structure will behave initially as uncracked for further load. Therefore, it is reasonable to consider a "concrete contribution" V_c for prestressed containments.

Figure E-5[30] demonstrates the difference in shear strength between initially uncracked and initially cracked interface shear test specimens.[31] An additional strength of approximately 250 psi or $4.0\sqrt{f_c'}$ is observed in the uncracked specimens. However, as discussed in the introduction of this report, interface shear test specimens model a relatively artificial situation.

The wall panel specimens tested under combined biaxial and shear stresses model the behavior more realistically. Of the panel test programs cited in the Introduction, the initially uncracked specimens tested at the University of Toronto[18] are applicable to behavior of a prestressed containment.

Figure E-6 shows the principle tensile stress at cracking, f_{cr}', in the Toronto specimens. Except for two specimens, PV2 and PV24,

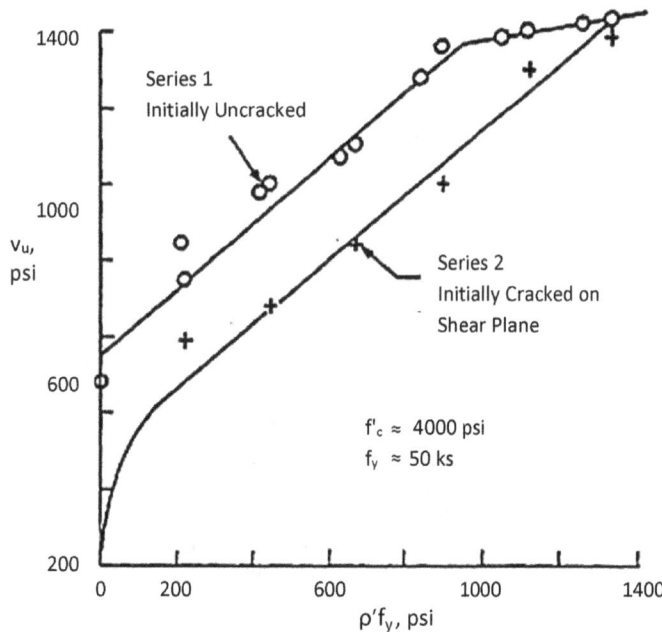

FIGURE E-5 VARIATION OF SHEAR STRENGTH WITH REINFORCEMENT PARAMETER ρf_y WITH AND WITHOUT A CRACK ALONG THE SHEAR PLANE.

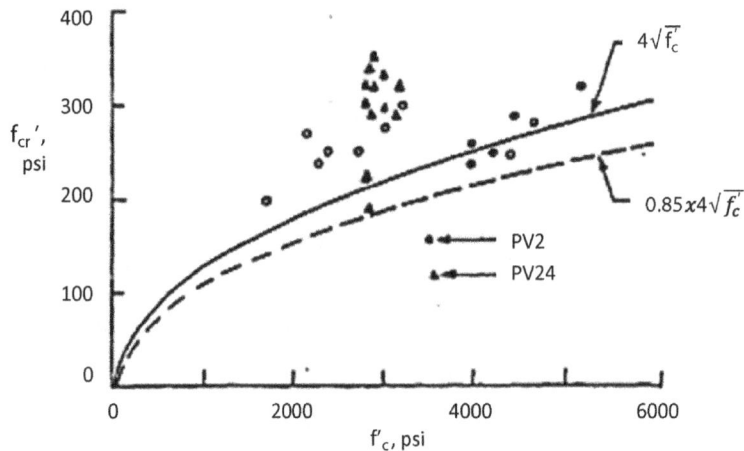

**FIGURE E-6   PRINCIPLE TENSILE CRACKING STRESS VERSUS
CYLINDER STRENGTH.**

all cracking stresses are close to or above the line indicating $4\sqrt{f_c'}$. Specimen PV2 was precracked and specimen PV24 had inadequately consolidated concrete. These test results confirm the diagonal cracking criteria in the ACI Building Code.[5]

The criteria for shear in prestressed concrete members contained in Section 11.4.2.2 of the ACI Building Code[5] allows a principle tensile stress of $4\sqrt{f_c'}$ in the web of the members. Therefore, it is recommended that initially a principle tensile stress of $4\sqrt{f_c'}$ be carried by the concrete in prestressed containments. This corresponds to following "concrete contribution" derived from Mohr's circle:

$$V_c = 4\sqrt{f_c'}\,bt\sqrt{1+\frac{f_m+f_h}{4\sqrt{f_c'}}+\frac{f_m+f_h}{\left(4\sqrt{f_c'}\right)^2}} \qquad (5)$$

where fm and f_h are positive for compression.

No additional reinforcement for shear reinforcement is required if V_u is less than $0.85\,V_c$. The 0.85 factor is a strength reduction normally associated with shear. If the shear load V_u is greater than $0.85\,V_c$ in Equation (5) then the concrete should be considered cracked with no "concrete contribution." The entire shear should be resisted by reinforcement according to Equations (3) and (4).

Maximum Strength

For reinforced concrete containments with orthogonal steel providing part of the shear strength, the current code[9] limits shear strength V_u to $8\sqrt{f_c'}\,bt$ for Factored Loads.

Limits on maximum shear strength are stated in building codes[5,32] for three reasons.

(a) Prevent a diagonal crushing failure in the truss mechanism of shear transfer.

(b) Prevent a sliding shear failure (a local combined shear-crushing failure along a horizontal plane) in the shear friction mechanism of shear transfer.

(c) Prevent large, unsightly shear cracks at the service load level.

The ACI 318 Building Code,[5] limits V_s to $8\sqrt{f_c'}\,bt$, which then limits V_u to about 10 to $12\sqrt{f_c'}\,bt$. These limits appear for crack control at sustained service load levels for non-prestressed beams with Grade 60 reinforcement.[33] Since tangential shear for sustained

service loads is negligible, crack control at service loads should not be a governing factor for containments. Strength and deformations at factored loads should govern. In general, without longitudinal or transverse steel yielding, the diagonal compression strength or sliding shear strength in reinforced containments should be significantly higher than $8\sqrt{f_c'}\,bt$.

For diagonal compression crushing, the CEB-FIP Model Code[32] allows $V_u = 0.3f_c'bt$ for orthogonal steel arrangements and up to $V_u = 0.45f_c'bt$ if diagonal shear reinforcement at $45°$ is used.

For sliding shear, Mattock[34] recommended a limit of $V_u = 0.3f_c'bt$ based on monotonically loaded monolithic push-off specimens and composite specimens with good bond between castings. This limit was reduced to $V_u = 0.24f_c'bt$ for reversing loads. Using large scale specimens similar to those tested by Mattock, Aoyagi[22] derived a "balanced" reinforcement ratio corresponding to a sliding shear strength limit of $0.27f_c'bt$.

The Japanese had proposed a shear strength limit of $0.18f_c'bt$.[19] However, this limit is based on test results of specimens with yielding horizontal reinforcement and significant shear distortions occurring prior to a shear failure.[20] A more recent proposed Japanese design criteria[28] is based on testing of full cylindrical models varying in size up to a 1/8 scale model.[23,24,27] These models exhibited maximum shear stress ranging from $19.6\sqrt{f_c'}$ to $22\sqrt{f_c'}$. Using a factor of safety of 1.5, new maximum shear strength limit of $13.2\sqrt{f_c'}\,bt$ is recommended by the Japanese, as shown in Figure E-4. For $f_c' = 4000$ psi, this limit is equal to $0.21f_c'bt$.

Panel specimens tested in the experimental programs conducted by CTL[14,15] and Cornell[16,17] all lost load capacity by yielding of reinforcement across diagonal cracks. Therefore, these data cannot be used to establish a limit on maximum strength. However, specimens tested by Vecchio and Collins[18] contained relatively high reinforcement ratios. Therefore, concrete crushing or sliding shear failures were the observed failure modes in most of the specimens.

Results indicated that concrete shear strength decreases as transverse and longitudinal tensile strains increased. With both transverse and longitudinal strains at zero, shear strength was $0.47f_c'bt$. However, with both transverse and longitudinal strain at 0.002, (typical yield strain for reinforcement) shear strength was $0.30f_c'bt$. The presence of biaxial tension and reversing shear load reduced shear strength to $0.25f_c'bt$. With a strength reduction factor of 0.85 to account for uncertainty normally associated with shear,[5] shear strength would be $0.21f_c'bt$.

Based on review of the available test data, it is recommended by the Task Group on Shear that maximum shear strength for factored loads V_u be limited to $0.2f_c'bt$ when orthogonal reinforcement is used to resist shear loads without inclined reinforcement present, i.e.,

$$V_{so} \leq 0.2f_c'bt$$

$$V_{so} = V_u - 0.9f_y A_{si} \tag{6}$$

It should be noted that $0.2f_c'$ is near the maximum tangential shear stress that might ever be expected in a containment. However, if additional strength is needed, inclined reinforcement can be used to increase the maximum shear strength.

When only orthogonal reinforcement is present, all the diagonal compressive stresses of the truss mechanism of shear transfer are resisted by concrete. Use of inclined reinforcement provides steel to resist a significant portion of the diagonal compression. When a symmetrical pattern of diagonal reinforcement is present, the strength of the inclined steel in compression balances the strength of inclined steel in tension.

Because of this balance of strength, it might be argued that the shear strength of a containment reinforced with inclined steel should only be limited by the amount of reinforcement that can be placed practically in the walls. However, compatibility must also be considered. The strain associated with yield of reinforcement of 60,000 psi is approximately 0.002. This is a very high compressive strain for concrete that is in tension in the orthogonal direction.

In order to evaluate the relationship between maximum concrete compressive stress and the amounts of orthogonal and diagonal reinforcement, a series of analyses of membrane elements were carried out. These analyses were made to evaluate parameters affecting the maximum shear stress and compressive stress corresponding to full yield of reinforcement in the element. The variables included orthogonal and inclined reinforcement ratios, ρ_m, ρ_h, and ρ_i, concrete strength, f_c', and concrete strain at peak stress, ε_o.

Analyses were made using equations of equilibrium and compatibility formulated by Duchon.[35] The equations were modified to account for yielding of the orthogonal reinforcement. The following nonlinear concrete stress-strain relationship based on panel tests at the University of Toronto [36] was used:

$$f_d = \frac{f_c'}{\beta}\left[2\frac{\varepsilon_c}{\varepsilon_o} - \left(\frac{\varepsilon_c}{\varepsilon_o}\right)^2\right] \tag{7}$$

$$\text{where} \quad \beta = 0.8 + 0.34\,\varepsilon_l/\varepsilon_o \tag{8}$$

Equation 7 is a relationship for the effective strength of the concrete in diagonal compression, f_d, as a function of the principal tensile strain ε_l. Effective concrete strength f_d, of the compression struts decreases as the tensile strain in the reinforcement running perpendicular through the strut increases. Solution was obtained using an interactive technique with an effective secant modulus for the concrete.

In the series of analyses, the orthogonal reinforcement was varied to represent designs with V_{so} ranging from 0 to $0.2f_c'bt$. With $V_{so} = 0$, the orthogonal reinforcement is designed to resist only membrane forces from pressurization. With $V_{so} = 0.2f_c'bt$, the orthogonal reinforcement was designed to resist normal forces from pressurization plus shear forces. $V_{so} = 0.2f_c'bt$ corresponds to the recommended maximum allowable design shear discussed in the preceding section for a containment with only orthogonal reinforcement.

At a particular level of orthogonal reinforcement, the amount of inclined reinforcement was increased in the analyses until a concrete crushing failure was calculated to occur prior to general yield of the reinforcement, i.e., $\sigma_\pi = f_d$.

Analyses were carried out with two different peak concrete strains, $\varepsilon_o = 0.0015$ and $\varepsilon_o = 0.002$. Varying affects the stiffness of the concrete and thereby affects the relative amounts of stress carried by the concrete and steel in compression.

Results of the analyses are shown in Figures E-7 through E-9. Figure E-7 shows the relationship between principle compressive stress σ_π and maximum shear for analyses made with $\varepsilon_o = 0.0015$. A diagonal compression crushing was calculated to occur at $\dfrac{\sigma_\pi}{f_c'} = 0.48$.

The maximum shear force at which crushing occurred is dependent on the amount of orthogonal reinforcement allocated to resist shear. Maximum shear strength decreases as V_{so} increases.

Figure E-8 presents data similar to Figure E-7 but with the principal stress normalized by f_d. In these plots σ_π/f_d represents a calculated diagonal compression crushing. The dashed line in Figure E-8

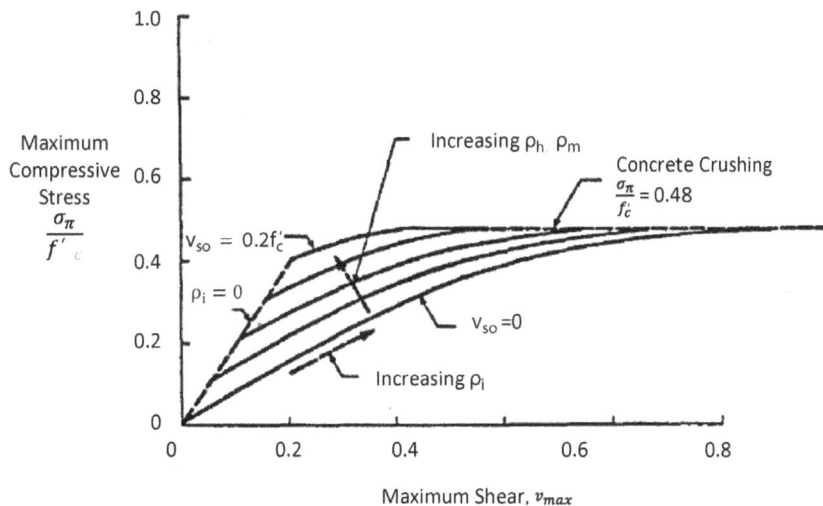

FIGURE E-7   MAXIMUM COMPRESSIVE STRESS (NORMALIZED BY f_c') VERSUS MAXIMUM SHEAR WITH VARYING AMOUNTS OF ORTHOGONAL AND DIAGONAL REINFORCEMENT USING $\varepsilon_0 = 0.0015$.

represents a conservative design limit for the concrete compressive stresses. It is drawn at $\sigma_\pi/f_d = 0.72$ which includes a strength reduction factor, $\phi = 0.85$, normally associated with shear, multiplied by another reduction of 0.85 to account for effects of load reversals.

Maximum shear strength as a function of V_{so} determined from the dashed line in Figure E-8 is shown by a dashed line in Figure E-9 along with a similar line determined for analyses with $\varepsilon_0 = 0.002$. These dashed lines in Figure E-9 demonstrate the relationship

between maximum shear strength and concrete stiffness. When inclined reinforcement is present, maximum strength decreases as concrete stiffness increases.

The solid line in Figure E-9 corresponds to the following recommended design limit for shear with or without diagonal reinforcement:

$$V_u \leq 0.4f_c'bt - V_{so} \tag{9}$$

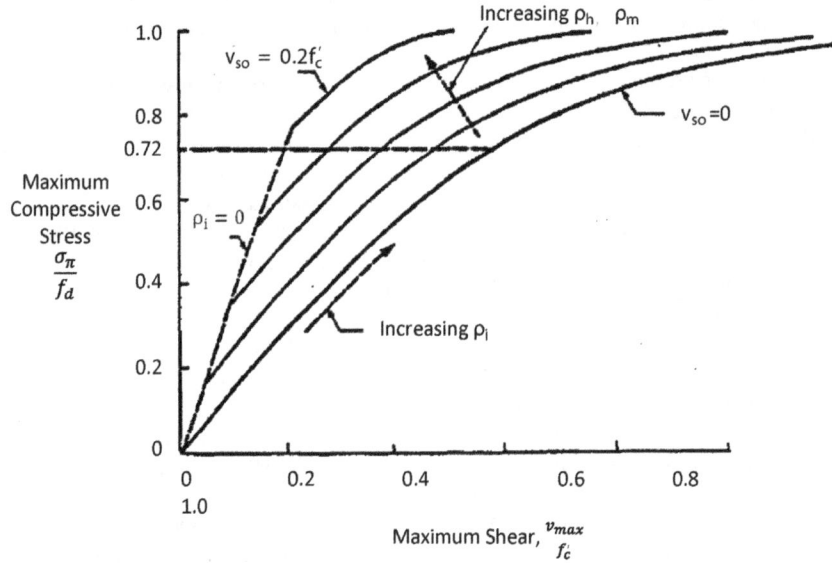

FIGURE E-8 MAXIMUM COMPRESSIVE STRESS (NORMALIZED BY f_d) VERSUS MAXIMUM SHEAR WITH VARYING AMOUNTS OF ORTHOGONAL AND DIAGONAL REINFORCEMENT USING $\varepsilon_0 = 0.0015$.

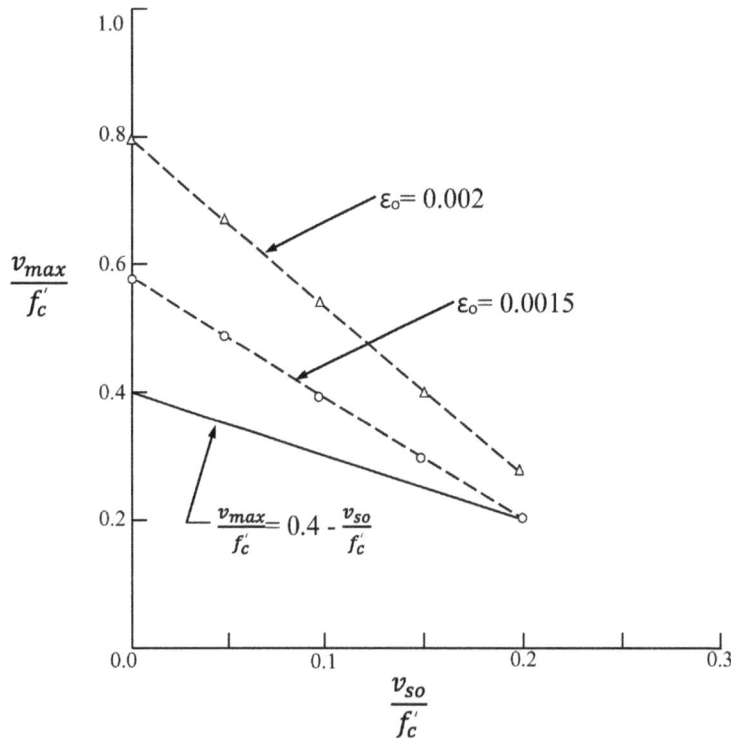

FIGURE E-9 MAXIMUM SHEAR STRESS VERSUS SHEAR STRESS RESISTED BY ORTHOGONAL REINFORCEMENT.

This limit is conservative even for very stiff concrete but will allow containment wall design for all practical situations.

Equation (9) implies that if $V_{so} = 0.2f'_c bt$, then $V_u = 0.2f'_c bt = V_{so}$. No inclined reinforcement can be added to increase strength because the orthogonal reinforcement has used up all available concrete strength in diagonal compression. If $V_{so} = 0$, then $V_u = V_{si} = 0.4f'_c bt$. In other words, a maximum shear corresponding to $0.4 f'_c$ can be obtained if shear forces are only carried by inclined reinforcement.

A combination of orthogonal and inclined reinforcement can be used to obtain intermediate strengths if $V_{so} < 0.2f'_c bt$. As an example, say $V_{so} = 0.1f'_c bt$, then $V_u = 0.3f'_c bt$ with $V_{si} = 0.2f'_c bt$.

4. DEFORMATION PROVISIONS

The strength provisions presented previously are intended to insure against loss of shear load capacity. However, because of the importance of leak-tightness integrity of the liner and interaction with attached equipment and piping, deformations should also be considered in design.

Results of testing concrete specimens with plane shear across a predefined crack,[1–4,6–8,10–13,22] panel specimens subjected to membrane shear,[14–18,21] and full cylindrical models[23–25,27] demonstrate that shear stiffness reduces significantly after cracking. As an example, in full cylindrical models tested with monotonic shear load by Bader and Krawinkler,[25] shear stiffness after cracking was 7.5% of the uncracked stiffness. Open cracks due to pressurization and abrasion from cyclic load will further reduce the shear stiffness.

Although this behavior of shear stiffness reduction has been known for some time, it is the recommendation of the Task Group on Shear that a statement regarding shear distortions be included in Section CC-3310 General (considerations), Containment Design Analysis Procedures of the ASME-ACI code.[9]

It is the opinion of the Task Group on Shear that the code should not prescribe how shear distortions should be considered other than to retain the current limit of $2\varepsilon_y$ for maximum strain in the reinforcement.

There are finite element methods available in current literature[26,37,38,39] for modeling cracked concrete. Finite element models have been used successfully to model deformation in full cylindrical models of containments.[24,26,27,40]

Shear distortion is expected to reach the sum of the strains in the meridional and hoop reinforcement.[41] The shear distortion can be limited approximately to the strain in the inclined reinforcement by providing inclined reinforcement to carry the entire tangential shear force, i.e., $V_{so} = 0$. The shear strain can also be limited by providing an excess amount of orthogonal reinforcement to carry tangential shear. Design considerations using this approach are suggested by Oesterle[42] based on observed shear deformations in panel tests. The criteria for acceptable deformations should be established from requirements of attached equipment and piping.

It should be noted that when a reducing shear stiffness model under reversing load is considered in dynamic analyses, the maximum shear stresses induced by seismic loading will probably be lower than the stresses normally calculated. Research is needed to develop simplified analytical procedures to efficiently incorporate shear deformations into design criteria.

5. SUMMARY

The purpose of this report is to present recommended design criteria for tangential shear based on available test data. The following is a summary of the recommendations by the Task Group on Shear:

1. The term V_c in the design equation of the current Code[9] should be replaced by V_{so}.
2. For reinforced concrete containments the "concrete contribution" V_c should be taken as zero.
3. Required area of orthogonal (hoop and meridional) reinforcement, with or without inclined reinforcement, provided for combined membrane and tangential shear strength shall be computed by:

$$A_{sh} + A_{si} = \frac{N_h + [N_{hl}^2 + V_u^2]^{1/2}}{0.9f_y} \quad (3)$$

$$A_{sm} + A_{si} = \frac{N_m + [N_{ml}^2 + V_u^2]^{1/2}}{0.9f_y} \quad (4)$$

4. For prestressed concrete containments V_C should be based on a principle tensile stress of $4\sqrt{f'_c}$ carried by the concrete. If V_U exceed $0.85V_c$, the entire shear should be resisted by reinforcement designed according to Equations (3) and (4).
5. Any combination of orthogonal and inclined reinforcement as required for strength according to Equations (3) and (4), and as required to control shear deformations may be used with the following limits on maximum shear force:

$$V_{so} \leq 0.2f'_c bt \quad (6)$$

$$\text{Where} \quad V_{so} = V_u - 0.9f_y A_{si}$$

$$V_u \leq -.4f'_c bt - V_{so} \quad (9)$$

6. A statement regarding consideration of shear distortions should be included in the Codes[9] statements on containment design and analysis procedures in Section CC-3310.
7. Further research should be conducted to develop simplified analytical procedures to efficiently incorporate shear deformations into design criteria.

Example calculations to determine required areas of reinforcement and to check maximum reinforcement strains with and without the use of inclined reinforcement are presented in Appendix B of this report.

6. NOTATIONS

A_{sh} = Area of bonded reinforcement in the hoop direction (in²/ft)

A_{sm} = Area of bonded reinforcement in the meridional direction (in²/ft)

A_{si} = Area of bonded reinforcement in one direction of inclined bars at 45° to horizontal (in²/ft along a line perpendicular to the direction of the bars). Inclined reinforcement shall be provided in both directions.

N_n and N_m = Membrane force in the hoop and meridional direction respectively due to pressure, prestress and dead load. N_h and N_m are positive when tension and negative when compression. The prestress force shall be the effective value.

N_{hl} and N_{ml} = Membrane force in the hoop and meridional direction respectively from lateral load such as earthquake, wind, or tornado loading. When considering earthquake loading, this force is based on the square root of the sum of the squares of the components of the two horizontal and vertical earthquakes. The force is always considered as positive and the units are k/ft.

V_c = Tangential shear strength provided by concrete.

V_s = Tangential shear strength provided by reinforcement = $V_{so} + V_{si}$

V_{so} = Tangential shear strength provided by orthogonal (hoop and meridional) reinforcement

V_{si} = Tangential shear strength provided by inclined reinforcement

V_u = The peak membrane tangential shear force resulting from lateral load such as earthquake, wind, or tornado loading. When considering earthquake loading, this force is based on the square root of the sum of the squares of the components of the two horizontal and vertical earthquakes. The shear force shall be considered as positive and the units are k/ft.

b = Unit length of section

f'_c = Compressive strength of standard 6×12-in. concrete cylinders

f_m = Concrete membrane stress in the meridional direction

f_h = Concrete membrane stress in the hoop direction

f_s = Maximum orthogonal reinforcement tensile stress from membrane forces N_h or N_m

f_y = Yield strength of reinforcement

t = Net wall thickness considering any reduction due to tendon ducts

v_{max} = Maximum observed shear stress in test specimens

v_{so} = V_{so}/bt = design shear stress for orthogonal reinforcement

v_u = V_u/bt = total design shear stress

ε_c = Strain in concrete

ε_o = Strain in concrete at peak stress

ε_y = Yield strain of reinforcement

v = Principle tensile strain in the membrane element

ρ = Lesser of ρ_h or ρ_m

ρ' = Effective reinforcement ratio for diagonal tension equilibrium of test specimens

ρ_h = Horizontal reinforcement ratio A_{sh}/bt

ρ_m = Vertical reinforcement A_{si}/bt ratio

ρ_i = Inclined or diagonal reinforcement ratio A_{si}/bt

σ_{II} = Principle compressive stress in the concrete

7. REFERENCES

1. Colley, B. E. and Humphrey, H. A., "Aggregate Interlock at Joints in Concrete Pavements," Research and Development Bulletin D124, Portland Cement Association, 1967, 18 pp.

2. Fenwick, R. C. and Paulay, T.O., "Mechanism of Shear Resistance in Concrete Beams," *Journal of the Structural Division*, ASCE, Vol. 94, No. ST 10, October, 1968.

3. Paulay, T. and Loeber. P.J., "Shear Transfer by Aggregate Interlock." ACI Special Publication SP42, *Shear in Reinforced Concrete, Vol. 1*, American Concrete Institute, 1974 pp. 1–16.

4. Houde, J. and Mirza, M. S., "Investigation of Shear Transfer Across Cracks by Aggregate Interlock," Research Report No. 71-06. Ecole Polytechnique de Montreal, Dept. of Gene Civil, Division de Structures, 1972.

5. ACI Committee 318, Building Code Requirements for Reinforced Concrete, ACI Standard 318-83, American Concrete Institute, Detroit, 1983, 111 pp.

6. White, R. N. and Holley, M. J., "Experimental Studies of Membrane Shear Transfer," *Journal of the Structural Division*. American Society of Civil Engineers, No. ST3. August 1972.

7. Laible, J. P., White, R.N., and Gergely, P., "Experimental Investigation of Seismic Shear Transfer Across Cracks in Nuclear Containment Vessels," ACI special Publication SP53, *Reinforced Concrete Structures in Seismic Zones*, American Concrete Institute, 1977, pp. 203–226.

8. White, R. N. and Gergely, P., "Shear Transfer in Thick Walled Reinforced Concrete Structures Under Seismic Loading," Department of Structural Engineering, Cornell University, Report No. 75-10.

9. "ASME Boiler and Pressure Vessel Code," Section III, Division 2, Code for Concrete Reactor Vessels and Containments, American Society of Mechanical Engineers, New York, 1983 Edition.

10. Jimenez, R., Gergeley. P., and White. R. N., "Shear Transfer Across Cracks in Reinforced Concrete," Report No. 78-4, Department of Structural Engineering, Cornell University, Ithaca, New York, August 1979.

11. Mattock, A. H., "Effect of Aggregate Type on Single Direction Shear Transfer Strength in Monolithic Concrete," Report SM74-2, University of Washington, August 1974, 34 pp.

12. Mattock, A. H., "The Shear Transfer Behavior of Cracked Monolithic Concrete Subjected to Cyclically Reversing Shear," Report SM74-4, University of Washington, November 1974, 31 pp.

13. Walraven, J. C., Vos., E., and Reinhardt, H. W., "Experiments on Shear Transfer in Cracks in Concrete, Part I: Description of Results," Report 5-79-3, Delft University of Technology 1979, 87 pp.

14. Oesterle, R. G. and Russell, H. G., "Shear Transfer in Large Scale Reinforced Concrete Containment Elements-Report No. 1," Report NUREG/CR-1374, U.S. Nuclear Regulatory Commission, Portland Cement Association, April 1980, 67 pp.

15. Oesterle, R. G. and Russell, H. G., "Shear Transfer in Large Scale Reinforced Concrete Containment Elements-Report No.2," Report to U.S. Nuclear Regulatory Commission, Portland Cement Association, July 1981, 85 pp.

16. Perdikaris, P. C., White, R. N., and Gerqely, P., "Strength and Stiffness of Tensioned Reinforced Concrete Panels Subjected to Membrane Shear: Two-Way Reinforcing," Report to NUREG/CR-1602, U.S. Nuclear Regulatory Commission, Cornell University, July 1980. 394 pp.

17. Conley, C. H., White., R. N., and Gergely, P., "Strength and Stiffness of Reinforced Concrete Panels Subjected to Membrane Shear: Two-Way and Four-Way Reinforcing," Report to NUREG/CR-2049, U.S. Nuclear Regulatory Commission, Cornell University, April 1981, 165 pp.

18. Vecchio, F. and Collins, M. P. "The Response of Reinforced Concrete to In-Plane Shear and Normal Stresses" Publication No. 82-03, University of Toronto, March 1982, 332 pp.

19. Ichikawa, K., Adyagi, Y., and Watanabe, Y., "Design Concepts of Concrete Containment Vessels for Shear and Thermal Stresses," Paper J4/1, Transactions of the 5th Conference on Structural Mechanics in Reactor Technology, Berlin, 1979.

20. Uchida, T., Ohmori, N., Takahashi, T., Watanabe, S., Abe, H., and Adyagi, Y., :Behavior of Reinforced Concrete Containment Models Under the Combined Action of Internal Pressure and Lateral Force," Paper J4/4, Transactions of the 5th Conference of Structural Mechanics in Reactor Technology, Berlin, 1979.

21. Aoyagi, Y., Ohmori, S., and Yamada, K., "Strength and Deformation Characteristics of Orthogonally Reinforced Concrete Containment Models Subjected to Lateral Forces."Paper J4/5, Transactions of the 6th Conference on Structural Mechanics in Reactor Technology, Paris, 1981.

22. Aoyagi. Y., Sakamoto, S., Takao, K., and Yamasaki, A., "Full Scale Model Push-Off Test of Reinforced Concrete Block With 51 mm Dia. Deformed Steel Bars," Paper J4/7, Transactions of the 6th Conference on Structural Mechanics in Reactor Technology, Paris, 1981.

23. Osaki, Y, Kobayashi, M., Takega, T., Yamaguchi, T., Yoshizaki, and Sugano, S., "Shear Strength Tests of Prestressed Concrete Containment Vessels," Paper J4/3, Transaction of the 6th Conference on Structural Mechanics in Reactor Technology, Paris 1981.

24. Ogaki, Y., Kobayash, M., Takeda, T., Yamaguchi, T., and Yoshioka, K., "Horizontal Loading Tests on Large-Scale Moore if Prestressed Concrete Containment Vessel," Paper J4/2, Transactions of the 6th Conference on Structural Mechanics in Reactor Technology, Paris 1981.

25. Bader, M. and Krawinkler, "Shear Transfer in Thick-Walls Reinforced Concrete Cylinders." Paper J3/7, Transaction of the 6th Conference on Structural Mechanics in Reactor Technology, Paris 1981.

26. Miyashita, T. and Sozen, M. A., "Nonlinear Analysis of Reinforced-Concrete Containment Vessel Using Shear Transfer Stiffness of Cracked Elements," Paper J3/12, Transaction of the 6th Conference on Structural Mechanics in Reactor Technology, Paris 1981.

27. Adyagi, A., Ohnuma, H., Ichikawa, K., and Isobata. O., "Test of a PCCV Under Load Combination of LOCA and Earthquake," Paper J4/9, Transaction of the 6th Conference on Structural Mechanics in Reactor Technology, Paris 1981.

28. Ohsaki, Y., IBE, Y., and Adyagi, Y., "Drafted Japanese Design Criteria for Concrete Containment," Paper J1/2, Transaction of the 6th Conference on Structural Mechanics in Reactor Technology, Paris 1981.

29. ACl-ASME Technical Committee on Concrete Pressure Components for Nuclear Service, Item No. JC 7-5, Code Case N-250, Alternate Rules for Design of Tangential Shear Forces for Section III, Division 2, Class CC, 1980.

30. Joint ASCE-ACI Task Committee 426, "The Shear Strength of Reinforced Concrete Members," ASCE Journal, Structural Division, ST6, June 1973, pp. 1091–1186.

31. Hofbeck, J. A., Ibrahim, I.A. and Mattock, H.H., "Shear Transfer in Reinforced Concrete," ACI Journal, Vol. 66, No. 2, February 1969, pp. 119–128.

32. Comite Euro-International Du Beton, "Shear and Torsion," CEB-FIP Model Code for Concrete Structures, Bulletin D'Information No. 126, Paris, June 1978.

33. Collin, M. P. and Mitchell, D., "Shear and Torsion Design of Prestressed and Non-Prestressed Concrete Beams," Journal,

Prestressed Concrete Institute, Vol. 25, No.5, Sept./Oct. 1980, pp. 32–100.

34. Mattock. A. H., "Shear Transfer Under Cyclically Reversing Loading Across an Interface Between Concrete Cast at Different Times," Report SM 77-1, University of Washington, Seattle, June 1977, 26 pp.

35. Duchon, N. B., "Analysis of Reinforced Concrete Membrane Subject to Tension and Shear," ACI Journal. Proceeding Vol. 69, No. 9, September 1972, pp. 578-583.

36. Private Communications Between M. Collins and R. Oesterle. Information is included in a paper that has been accepted for publication in ACI Journal.

37. ASCE Task Committee on Finite Element Analysis of Reinforced Concrete Structures, Finite Element Analysis of Reinforced Concrete Structures, American Society of Civil Engineers, 1982, 545 pp.

38. Calvo, J., Buyukozturk, O., and Connor, J. J., "Design of Reinforced Concrete Containment Wall Elements Under Combined Action of Shear and Tension," Report to U.S. Nuclear Regulatory Commission, NUREG/CR 3157, Massachusetts Institute of Technology, February 1983, 211 pp.

39. Gupta, A. K., "Design of Membrane Reinforcement for Concrete Nuclear Containments" North Carolina State University, January 1982, 25 pp.

40. Conley, C. H., White, R. N., and Gergely, P., "Analysis of Reinforced Concrete Containment Vessels with Nonlinear Shearing Stiffness, " Report NUREG/CR-3255. U.S. Nuclear Regulatory Commission, Cornell University, April 1983, 98 pp.

41. P. Shummugavel, "Analysis and Design of Concrete Containments for Tangential Shear Loads," Paper No. J5/2, 7th International Conference on Structural Mechanics in Reactor Technology, August 1983.

42. Oesterle. R. G., "Tangential Shear Design in Reinforced Concrete Containments: Research Results and Applications," Transactions of the 7th International Conference on Structural Mechanics in Reactor Technology, Panel Session JK-P, August 1983.

APPENDIX A

Force Distribution

A concrete containment shell is generally a vertical thin-walled cantilever structure with a circular cross section. Tangential shear forces are in the plane of the containment shell resulting from lateral loading such as wind or seismic loads.

Wind Load

The lateral wind load causes an overall moment M and shear V, both of which are varying along the height of the containment (Figure A1). The resulting stresses at an elevation of the containment are shown in Figure A2 corresponding to an uncracked elastic condition. The maximum meridional force N_{VW} occurs at the outermost fiber of the cross section while the maximum tangential shear force V_{UW} occurs at the centerline of the cross section. The maximum forces are given by:

$$N_{vw} = \frac{M}{\pi r^2}$$

$$N_{vw} = \frac{M}{\pi r^2}$$

where
  r = mean radius of the containment cross-section

The forces at any location along the circumference of the containment are expressed as:

$$N_\theta = N_{vw} \cos \theta$$

$$V_\theta = V_{uw} \sin \theta$$

where
  θ = angle from the direction of wind (Figure A2).

The maximum value of $(N_\theta + V_\theta)$ occurs at θ_1 which can be derived as:

$$(N_\theta + V_\theta)_{max} = \sqrt{N_{vw}^2 + V_{uw}^2}$$

$$\theta_1 = Tan^{-1}\left(\frac{V_{uw}}{N_{vw}}\right)$$

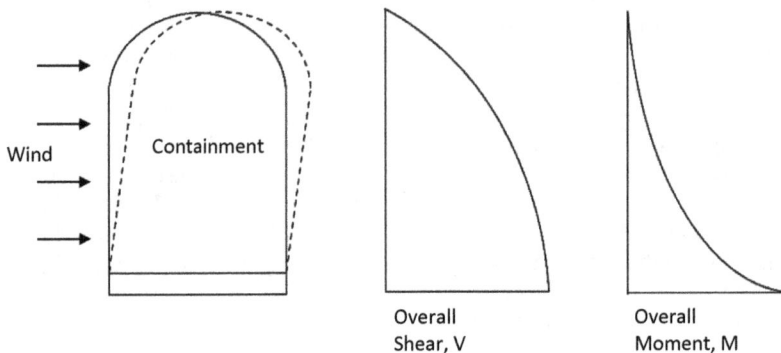

FIGURE A1 SHEAR AND MOMENT ON CONTAINMENT FROM WIND LOADING.

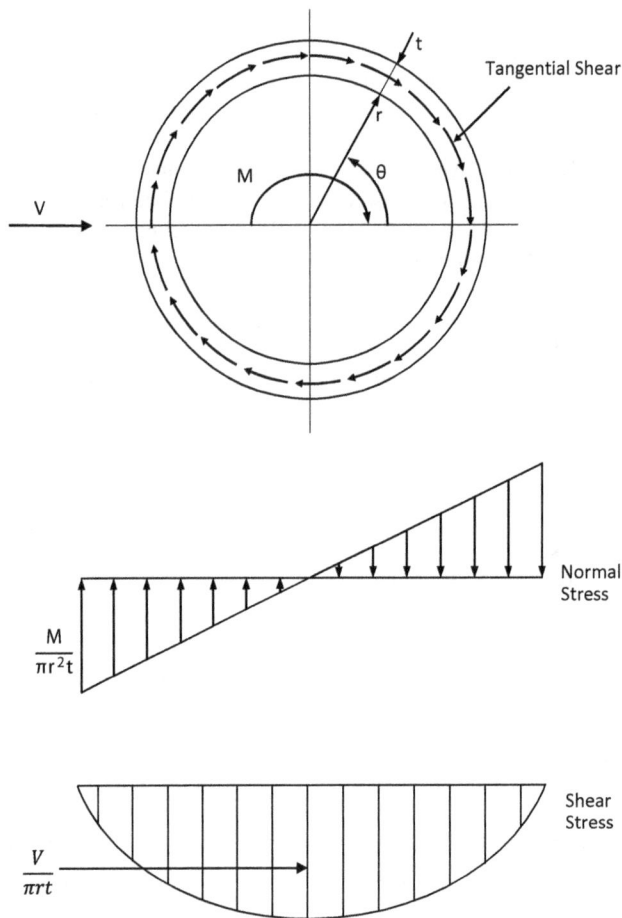

FIGURE A2 DISTRIBUTION OF STRESSES FROM WIND LOAD.

Earthquake Load

An earthquake has three orthogonal (two horizontal and one vertical) components which cause the following distributions of stresses corresponding to an uncracked elastic condition as shown in Figure A3. The internal forces at a location defined by angle θ are expressed as:

$$N_\theta = \pm E_h \cos \theta; \pm E_h \sin \theta; \pm E_v$$

$$V_\theta = \pm T \sin \theta; \pm T \cos \theta; 0$$

where

$E_h = \dfrac{M}{\pi r^2}$ = maximum meridional force from a horizontal component of the earthquake

$E_v = \dfrac{N}{2\pi r}$ = meridional force from the vertical component

$T = \dfrac{V}{\pi r}$ = maximum tangential shear from a horizontal component

N = overall meridional force from the vertical component

M, V = overall moment and shear from a horizontal component

It should be noted from Figure A3 that all locations along the circumference, either the meridional force N_θ or the shear V_θ from a horizontal earthquake component has an opposite sign compared to those from the other horizontal component.

The sum of the meridional and shear forces at a location is expressed as:

$$(N_\theta + V_\theta) = \pm(E_h \cos \theta + T \sin \theta);$$
$$\pm(E_h \sin \theta - T \cos \theta); \pm E_v$$

Combining the responses from the three components by the square-root-of-the-sum-of-squares (SRSS) method,

$$(N_\theta + V_\theta) = \pm[E_h^2(\cos^2 \theta + \sin^2 \theta)$$
$$+ T^2(\sin^2 \theta + \cos^2 \theta) + E_v^2 + 2E_h T \cos\theta \sin\theta$$
$$- 2E_h T \sin\theta \cos\theta]^{1/2}$$
$$= \pm[E_h^2 + E_v^2 + T^2]^{1/2}$$

$$(N_\theta + V_\theta) = \pm\sqrt{N_{ve}^2 + V_{ue}^2}$$

where

$$N_{ve}^2 = E_h^2 + E_v^2$$

$$V_{ue} = T$$

Thus, the total response $(N_\theta + V_\theta)$ from the three earthquake components are the same at all locations, i.e., independent of angle θ.

APPENDIX B

EXAMPLE CALCULATIONS FOR DESIGN FOR TANGENTIAL SHEAR—MEMBRANE REGION
Design Parameters: 0.6g SSE, P_a = 52 psig

$f_c' = 3$ ksi	$f_y = 60$ ksi	$b = 12$ in
$E_c = 3150$ ksi	$E_s = 29.000$ ksi	$t = 53.625$

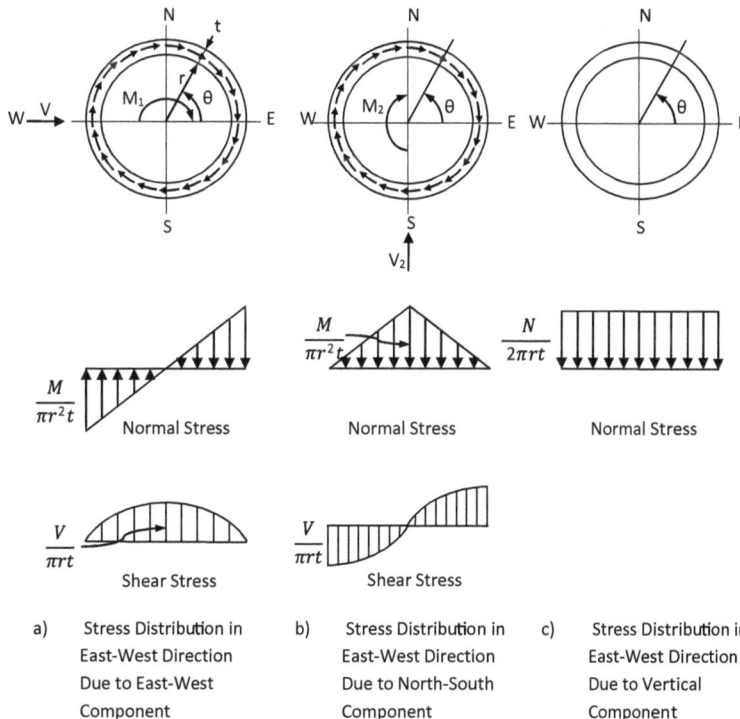

FIGURE A3 DISTRIBUTION OF STRESSES FROM EARTHQUAKE LOADS.

Load Combination: $D + P_a + E_{ss}$

$$N_m = 116 \text{ k/ft} \qquad N_h = 480 \text{ k/ft}$$
$$N_{ml} = 504 \text{ k/ft} \qquad N_{hl} = 17 \text{ k/ft} \qquad v = 324 \text{ k/ft}$$

$$A_{sh} + A_{si} = \frac{N_h + (N_{hl}^2 + v_u^2)^{1/2}}{0.9f_y}$$
$$= \frac{480 + (17^2 + 324^2)^{1/2}}{0.9 \times 60} = 15.0 \text{ in.}^2/\text{ft} \qquad (3)$$

$$A_{sm} + A_{si} = \frac{N_m + (N_{ml}^2 + V_u^2)^{1/2}}{0.9f_y}$$
$$= \frac{116 + (504^2 + 324^2)^{1/2}}{0.9 \times 60} = 13.25 \text{ in.}^2/\text{ft} \qquad (4)$$

$$0.2f_c'bt = 0.2 \times 3 \times 12 \times 53.625 = 386 \text{ k/ft}$$

V_{so} = Tangential shear provided by orthogonal rebars

$$= (V_u - 0.9A_{si}) \le 0.2f_c'; bt \qquad (6)$$

$$Vu = 324 \text{ k/ft} < 0.2f_c'bt = 386 \text{ k/ft}$$

No inclined shear rebar is required.

Load Combination: $D + 1.25\, P_a + 1.25\, E_o$

$$N_m = 179 \text{ k/ft} \qquad N_h = 598 \text{ k/ft}$$
$$N_{ml} = 400 \text{ k/ft} \qquad N_{hl} = 13 \text{ k/ft} \qquad V = 255 \text{ k/ft}$$

$$A_{sh} + A_{si} = \frac{N_h + (N_{hl}^2 + V_u^2)^{1/2}}{0.9f_y}$$
$$= \frac{598 + (13^2 + 255^2)^{1/2}}{0.9 \times 60} = 15.8 \text{ in.}^2/\text{ft} \qquad (3)$$

$$A_{sm} + A_{si} = \frac{N_m + (N_{ml}^2 + V_u^2)^{1/2}}{0.9f_y}$$
$$= \frac{179 + (400^2 + 255^2)^{1/2}}{0.9 \times 60} = 12.0 \text{ in.}^2/\text{ft} \qquad (4)$$

Considering both load combinations, rebars required are as follows:

$$A_{sh} + A_{si} = 15.8 \text{ in.}^2/\text{ft}$$

$$A_{sm} + A_{si} = 13.25 \text{ in.}^2/\text{ft}$$

Provide the following rebars:

Example (a): Without Inclined Seismic Rebars

$$A_{sh} = 16.25 \text{ in.}^2/\text{ft} \quad A_{sm} = 13.50 \text{ in.}^2/\text{ft} \quad A_{si} = 0$$

Example (b): With Inclined Seismic Rebars

$$A_{sh} = 16.25 - 3.2 \qquad A_{sm} = 13.5 - 3.2$$
$$A_{sh} = 13.05 \text{ in.}^2/\text{ft} \qquad A_{sm} = 10.3 \text{ in.}^2/\text{ft}$$
$$A_{si} = A_{s3} = A_{s4} = 3.2 \text{ in.}^2/\text{ft}$$

Required reinforcement areas are determined based on a SRSS of normal and shear forces resulting from earthquake loading as discussed in Appendix A. To determine membrane forces to use in calculations for strain compatibility to check maximum reinforcement strain, make the following adjustments to the forces:

a.   Load Combination: $D + P_a + E_{ss}$

$$N_h' = N_h + (N_{hl}^2 + V_u^2)^{1/2} - V_u = 480 + (17^2 + 324^2)^{1/2} - 324 = 481 \text{k/ft}$$

$$N_m' = N_m + (N_{ml}^2 + V_u^2)^{1/2} - V_u = 116 + (504^2 + 324^2)^{1/2} - 324 = 391 \text{k/ft}$$

$$V_u = 324 \text{k/ft}$$

b.   Load Combination: $D + 1.25P_a + 1.25E_o$

$$N_h' = 598 + (13^2 + 255^2)^{1/2} - 255 = 598 \text{ k/ft}$$

$$N_m' = 179 + (400^2 + 255^2)^{1/2} - 255 = 398 \text{ k/ft}$$

$$V_u = 255 \text{ k/ft}$$

Initial results indicate incline tension reinforcement will yield. Therefore, the calculations must be repeated with adjustments for effects of yielding of inclined seismic reinforcement:

Stresses in tensile diagonal rebars are as follows:

$$f_{s3}(D + P_a + E_{ss}) = 87.49 \text{ ksi} > 0.9f_y(54 \text{ ksi}), \varepsilon_{s3} > \varepsilon_y$$

$$f_{s3}(D + 1.25P_a + 1.25E_{ss}) = 84.95 \text{ksi} > 0.9f_y(54\text{ksi}), \varepsilon_{s3} > \varepsilon_y$$

Restrict f_{s3} to 54 ksi and neglect it for any further load – carrying purpose. Read just membrane axial forces and tangential shear neglecting tensile inclined rebar (A_{s3}) and forces associated with it.

Adjusted forces:
a.   Load Combination: $D + P_a + E_{ss}$

$$N_m'' = 391 - 54 \times 3.2 \times 0.5* = 304.6 \text{ k/ft}$$

$$N_m'' = 481 - 54 \times 3.2 \times 0.5* = 394.6 \text{ k/ft}$$

$$V_u'' = 324 - 54 \times 3.2 \times 0.5* = 237.6 \text{ k/ft}$$

b.   Load Combination: $D + 1.25P_a + 1.25E_O$

$$N_m'' = 398 - 54 \times 3.2 \times 0.5* = 311.6 \text{ k/ft}$$

$$N_h'' = 598 - 54 \times 3.2 \times 0.5* = 511.6 \text{ k/ft}$$

$$V_u'' = 255 - 54 \times 3.2 \times 0.5* = 168.6 \text{ k/ft}$$

---

* $\sin^2 45 = 0.5$

TABLE B-1 DESIGN FOR TANGENTIAL SHEAR—MEMBRANE REGION: SUMMARY OF INITIAL RESULTS*

Rebar Patterns	Load Comb.	Force System (k/ft)	Adjusted Forces for Compatibility	Rebars Req'd (in²/ft) $A_{sh}+A_{sl}$	$A_{sm}+A_{sl}$	Rebars Provided (in²/ft) A_{sh}	A_{sm}	A_{sl}	$A_{sh}+A_{sl}$	$A_{sm}+A_{sl}$	Computed Rebar Stresses (ksi) Merd. f_m	Hoop f_h	Seismic f_{s3}	f_{s4}	f_c (ksi)	θ**	$v_s=$ $f_c=$ $\sin\beta \times \cos\beta$ (ksi)	Remarks
Orthogonal Rebars Only	D + Pa + Ess	$N_m = 116$ $N_{ml} = 504$ $N_h = 480$ $N_{he} = 17$ $V_u = 324$	$N'_m = 391$ $N'_h = 481$ $V_u = 324$	15.00	13.25	16.25	13.50	0	16.25	13.50	52.47	49.96	0	0	−1.0	44.4°	0.5	$v_s = 0.50$ksi $<0.2f'_c$
	D + 1.25Pa + 1.25EQ	$N_m = 179$ $N_{ml} = 400$ $N_h = 598$ $N_{he} = 13$ $V_u = 255$	$N'_m = 398$ $N'_h = 598$ $V_u = 255$	15.80	12.00	16.25	13.50	0	16.25	13.50	48.90	52.1	0	0	−0.793	45.79°	0.397	$v_s = 0.397$ksi $<0.2f'_c$
Orthogonal and Diagonal Rebar	D + Pa + Ess	$N_m = 116$ $N_{ml} = 504$ $N_h = 480$ $N_{he} = 17$ $V_u = 324$	$N'_m = 391$ $N'_h = 481$ $V_u = 324$	15.00	13.25	13.05	10.30	3.2	16.25	13.50	41.96	40.54	87.49	−5.07	−0.547	44.53°	0.273	$f_{s3} = 87.49$ksi >54ksi limit f_{s3} to 54 ksi and read just the stresses
	D + 1.25Pa + 1.25EQ	$N_m = 179$ $N_{ml} = 400$ $N_h = 598$ $N_{he} = 13$ $V_u = 255$	$N'_m = 398$ $N'_h = 598$ $V_u = 255$	15.80	12.00	13.05	10.30	3.2	16.25	13.50	37.81	43.95	84.95	−3.19	−0.355	46.99°	0.177	$f_{s3} = 84.95$ksi >54ksi limit f_{s3} to 54 ksi and read just the stresses

* Calculations made using computer program to solve equations of equilibrium and compatibility from Reference 35.
** Angle between meridional and principal major axis.

TABLE B-2 DESIGN FOR TANGENTIAL SHEAR—MEMBRANE REGION: SUMMARY OF FINAL RESULTS*

Rebar Patterns	Load Comb.	Force System (k/ft)	Adjusted Forces for Compatibility	Rebars Req'd (in²/ft) $A_{sh}+A_{sl}$	$A_{sm}+A_{sl}$	Rebars Provided (in²/ft) A_{sh}	A_{sm}	A_{sl}	$A_{sh}+A_{sl}$	$A_{sm}+A_{sl}$	Merd. f_m	Hoop f_h	Seismic f_{s3}	f_{s4}	f_c (ksi)	θ**	$v_s=$ $f_c=$ Sin β x Cos β (ksi)	Remarks
Orthogonal Rebars Only	$D + P_a + E_{ss}$	$N_m = 116$ $N_{ml} = 504$ $N_h = 480$ $N_{he} = 17$ $V_u = 324$	$N'_m = 391$ $N'_h = 481$ $V_u = 324$	15.0	13.25	16.25	13.5	0	16.25	13.5	52.47	49.96	0	0	−1	44.4°	0.5	$v_s = 0.50$ksi $<0.2f'_c$ $\gamma = 0.00418$
	$D + 1.25P_a + E_{ss} + 1.25E_Q$	$N_m = 179$ $N_{ml} = 400$ $N_h = 598$ $N_{he} = 17$ $V_u = 255$	$N'_m = 398$ $N'_h = 598$ $V_u = 255$	15.8	12.00	16.25	13.5	0	16.25	13.5	48.9	52.10	0	0	−0.793	45.79°	0.397	$v_s = 0.392$ksi $<0.2f'_c$ $\gamma = 0.00338$
Orthogonal and Diagonal Rebar	$D + P_a + E_{ss}$	$N_m = 116$ $N_{ml} = 504$ $N_h = 480$ $N_{he} = 17$ $V_u = 324$	$N'_m = 391$ $N'_h = 481$ $V_u = 324$	15.0	13.25	13.05	10.3	3.2	16.25	13.5	52.12	48.99	54	−6.55	−0.707	44.21°	0.353	$\gamma = 0.00394$ $f_{s3} = 54$ksi $E_{s3} = \dfrac{108.69}{E_s}$
	$D + 1.25P_a + 1.25E_Q$	$N_m = 179$ $N_{ml} = 400$ $N_h = 598$ $N_{he} = 17$ $V_u = 255$	$N'_m = 398$ $N'_h = 598$ $V_u = 255$	15.8	12.00	13.05	10.3	3.2	16.25	13.5	47.34	51.7	54	−4.63	−0.503	46.75°	0.252	$\gamma = 0.00373$ $f_{s3} = 54$ksi $E_{s3} = \dfrac{103.61}{E_s}$

* Calculations made using computer program to solve equations of equilibrium and compatibility from Reference 35.

** Angle between meridional and principal major axis.

Final strains in tensile diagonal rebars

a.   Load Combination: D + P$_a$ + E$_{ss}$

$$\varepsilon_{s3} = \frac{f_m + f_h}{2E_s} + \frac{\gamma}{2} = \frac{54.12 + 48.99}{2 \times 29{,}000} + \frac{.00394}{2} = 0.00375$$

$$= \frac{108.68}{E_s} < 2\varepsilon_y \frac{120}{E_s}$$

b.   Load Combination: D + 1.25P$_a$ + 1.25E$_O$

$$\varepsilon_{s3} = \frac{f_m + f_h}{2E_s} + \frac{\gamma}{2} = \frac{47.34 + 51.70}{2 \times 29{,}000} + \frac{.00373}{2} = 0.00357$$

$$= \frac{103.61}{E_s} < 2\varepsilon_y \frac{120}{E_s}$$

Equilibrium Checks: A$_g$ = 643.5 in^2/ft,   A$_{sm}$ = 10.3 in^2/ft,
A$_{sh}$ = 13.05 in/ft,   A$_{si}$ = A$_{si}$ = 3.2 in^2/ft

a.   Load Combination: D + P$_a$ + E$_{ss}$ (ksi)

$$N'_m = 10.3 \times 52.12 + \frac{3.2}{2}(54 - 6.55) - 643.5 \times .707 \times \sin^2\beta = 391.5 \qquad (391\ k/ft)$$

$$N'_h = 13.05 \times 48.99 + \frac{3.2}{2}(54 - 6.55) - 643.5 \times .707 \times \cos^2\beta = 481.5 \qquad (481\ k/ft)$$

$$V'_u = \frac{3.2}{2} \times (54 + 6.55) + 643.5 \times .707 \times \sin\beta\cos\beta = 324 \qquad (324\ k/ft)$$

b.   Load Combination: D + 1.25P$_a$ + 1.25E$_O$
(β = 46.15°, f$_c$ = −0.503 ksi)

$$N'_m = 10.3 \times 47.34 + \frac{3.2}{2}(54 - 4.63) - 643.5 \times .503 \times \sin^2\beta = 398.3 \qquad (398\ k/ft)$$

$$N'_h = 13.05 \times 51.7 + \frac{3.2}{2}(54 - 4.63) - 643.5 \times .503 \times \cos^2\beta = 598.3 \qquad (598\ k/ft)$$

$$V'_u = \frac{3.2}{2} \times (54 + 4.63) + 643.5 \times .503 \times \sin\beta\cos\beta = 255.6 \qquad (255\ k/ft)$$

Appendix F—Tendon End Anchor Testing

[F-1] SUMMARY TESTING OF LARGE PRESTRESSING TENDON END ANCHOR REGIONS, THEODORE E. JOHNSON, BECHTEL POWER CORPORATION, SEPTEMBER 1973, PAPER 148 EXPERIENCE IN THE DESIGN, CONSTRUCTION, AND OPERATION OF PRESTRESSED CONCRETE PRESSURE VESSELS AND CONTAINMENTS FOR NUCLEAR REACTORS, UNIVERSITY OF YORK, SEPTEMBER 1975

Theodore E. Johnson, MS APPLIED MECHANICS, BSCE
Bechtel Power Corporation, San Francisco, California
International Conference on Experience in the Design, Construction and Operation of Prestressed Concrete Pressure Vessels and Containments for Nuclear Reactors
Sponsored by: The Institution of Mechanical Engineers, The Institution of Civil Engineers, The British Nuclear Energy Society
University of York, England, 8–12 September 1975

SUMMARY

The information presented in the paper is based on the testing documented in items: (a) and (b) below. Tests were performed on concrete end anchorage regions for prestressing tendons with ultimate strengths of approximately 8900 kN. One test structure simulated a full scale concrete containment buttress and the other two test specimens were concrete blocks. The behavior of the test structure and specimens, when subjected to loading, was monitored by strain gages and dial gages. The testing illustrated that all of the amounts of reinforcing tested should be acceptable for the end anchor zones of large tendons presently used in prestressed concrete containment structures.

(a) Topical Report BC-TOP-7, Full Scale Buttress Test for Prestressed Nuclear Containment Structures, T. Johnson, R. Marsh, September 1972, Bechtel Corporation, Reference 9.

(b) Topical Report BC-TOP-8, Tendon End Anchor Reinforcement Test, H. Franklin, T. Johnson, September 1972, Bechtel Corporation, Reference 10.

Appendix G—Liner Plate Design Example, Test Results and Studies

TABLE OF CONTENTS

1. SUMMARY

The report gives a thorough description of the liner plate together with figures illustrating all the applicable details. Problem areas and factors affecting the liner plate and anchorage system design are covered in great detail. The loads considered in the liner plate design are discussed and the approximate strains in the liner plate resulting from these loads are stated.

The occurrence of the loads and conditions imposed upon the liner plate in the analysis are highly improbable. The liner plate was subjected to the effects of prestress, concrete creep and shrinkage, dead load, maximum hypothetical earthquake, accident thermal gradients, and accident pressure. The accumulative total strain in the longitudinal direction was -1616 μ in/in, and -1788 μ in/in in the hoop direction. In order to have these strains convert to maximum stresses, it was assumed that the liner plate would have a uniaxial yield stress of at least 53.7 ksi. This resulted in a hoop stress of -74.5 ksi and a longitudinal stress of -71 ksi. The analysis then assumed that a panel with initial inward curvature would be adjacent to a panel with initial outward curvature. The panel with outward curvature was assumed to be 16% thicker than the panel with inward curvature. All of the preceding conditions lead to a very large and highly improbable membrane load of 21.6 kips/in. The form of the solution was to remove a 1″ strip of the liner plate in the hoop direction. At this point, the benefits of biaxial stiffening were entirely neglected. The panel with inward curvature was assumed to have a maximum inward deviation of 1/8″. The solution assumed that the concrete shell moves radially inward so that the membrane force of 21.6 kips/in. would exist. Since the panel with inward curvature acts primarily in bending and the panel with outward curvature acts in membrane compression, the anchor between these two panels will tend to displace toward the panel with inward curvature due to the difference in stiffness of the two panels. The movement is resisted by the panel with inward curvature together with the anchor embedded in concrete. Also, as the anchor displaces, the membrane load will tend to reduce. As this happens, other anchors tended to become unbalanced with respect to force and a method was used which considers the effects of all anchors. In solving the problem, test results were used which gave the load vs. displacement curves of both the concrete anchor and the panel with inward curvature, or bent plate. The data used for the bent plate did not consider any benefit from biaxial stiffening which will result in the real structure. The data used for the anchor did not consider the benefits gained by having concrete with a higher modulus of elasticity than that of the tests.

To simplify the solution, an imaginary force was calculated which considered the effects of all anchors acting upon the anchor under consideration so that only a single anchor needed to be analyzed. At this point, a membrane load was calculated to simulate the effects of internal pressure which tend to reduce the load on the anchor. This value was then subtracted from the previous imaginary force. Referring to Case II of Section 5.0, the liner plate anchorage system was then subjected to a self-relieving type of load with an initial force of 26.37 kips/in. The anchorage system was then allowed to displace tangentially until equilibrium was reached. Due to the fact that both the anchorage system and the bent plate went over yield, a plastic solution was used which considered the non-linear force vs. displacement relationship of both the bent plate and the anchor. After reaching equilibrium, it was found that the anchor had displaced .0318 inches and had developed a force of 4.41 kips/in. The energy used up in the anchor in obtaining equilibrium was calculated to be .1039 kips-in/in. The total energy available in the anchor was defined as .541 kips-in/in. This value is the lowest total energy observed in the testing of the anchorage system. The safety factor pertaining to the effects of tangential loads was the total energy available divided by the energy up in obtaining equilibrium. In the previous case, the safety factor is 5.2. Since the force on the anchor is 4.4 kips/in, and the embedment length is 3″, the concrete will have an average stress of 1470 psi; but in reality, as the anchor starts to displace, the stresses on the inner face of the concrete shell will increase rapidly until yield is reached, and the stresses further away from the liner plate will increase until a balanced condition is obtained. Due to the fact that the anchor is required to displace .0318 in. as predicted by the analysis, it is expected that there will be some local crushing of the concrete within the first 1/2 inch of the concrete. In the report other cases have been analyzed with various concrete properties.

As the panel with inward curvature deforms inward due to anchor movement, the panel will be subjected to flexure, and moments will develop at the anchor attachment points. The flexural stress or strain from these moments is also self-relieving since the panel will tend to have limited deformation. As an upper limit, a plastic hinge will form. The weld will then be subjected to both tangential shear and moment. The amount of rotation under accident conditions is not of the magnitude of the rotation found during testing. During the tests, summarized

in Appendix B, the anchor weld proved its ability to withstand both pure shear strain in conjunction with rotational strain due to bending.

The anchor to liner plate attachment weld will also be subjected to strains in the longitudinal direction. These longitudinal strains may approach yield, but relative to the strains occurring in the tangential direction, they are of a very small magnitude and when combined with all other strains, this will only produce a maximum principal shear strains lightly greater than the tangential shear strain.

The anchorage system may at times be subjected to axial pullout loads. These pullout loads will have a negligible effect on the weld. The area of concern is the anchorage into the concrete. The tensile load will be resisted by bond on one side of the anchor and shear on the other, but even without bond and shear, the pullout would have to overcome a large friction force imposed by shear load. Since the expected maximum pull out force is approximately 310 lbs./in., radial pullout should not be a problem due to the 1500 lbs./in. capacity of the system. The possibility of longitudinal buckling of the liner plate and anchorage system was considered in the analysis and it was found that in the worst condition, a force of 560 lbs./in. was needed to prevent longitudinal buckling. Since the system should have a capacity of 1500 lbs./in., longitudinal buckling is not considered to be a problem.

The conditions and resulting loads and displacements mentioned previously, will only occur at very isolated locations in the liner plate, if at all due to the improbability of having a simultaneous occurrence of all imposed conditions. The forces and displacements found during operation are very small relative to those occurring during an accident condition. Since the accident condition is a one-time occurrence, and the nature of the loads imposed upon the liner plate are self-limiting, it is only reasonable to allow the occurrence of large strains in the anchorage system relative to yield.

In conclusion, the report verifies through the use of analysis and test results that a failure of the liner plate anchorage system will not occur under an accident condition in conjunction with an earthquake even when all other worst conditions are assumed to occur simultaneously.

2. DESCRIPTION OF PROBLEM

2.1 General

Prestressed concrete and reinforced concrete reactor buildings which house Pressurized Water Reactors presently use carbon steel liner plates. The liner plate functions as a gas-barrier to prevent uncontrolled release of fission products from the reactor buildings during operation and also in the unlikely event of an accident, which would release large amounts of fission products into the reactor building. In the event of such an accident that is considered credible, the temperature inside the reactor building is predicted to rise to approximately 280° with a pressure of about 55 psig. At this time, the liner plate is called upon to limit leakage to approximately .1 to .2% per 24 hours, of the contained weight of gas so that fission product leakage is also limited. Due to the required high leak-tight integrity of the liner plate, specific care must be given to its design, fabrication, and erection. In general, the following report will discuss the liner plate details; factors affecting the liner plate such as load conditions; detailed theoretical analyses and sample calculations; applicable test results; and the results of parametric studies.

2.2 LINER PLATE DETAILS

(See Figure G-1 for details of the liner plate)

2.2.1 Cylinder

The cylinder is fabricated from 1/4″ ASTM A-442 plate together with ASTM A-36 longitudinal stiffener angles welded to the liner and also horizontal ASTM A-36 channels, angles, and flat bars attached to the liner plate and the longitudinal angles. The rolled structural shapes and flat bars both stiffen the liner plate during erection and concrete placement and anchor the liner plate to the hardened concrete.

2.2.2 Dome

The dome is similar to the cylinder except it only has angle stiffeners which also act as anchors; the attached angles are oriented approximately in the hoop direction.

2.2.3 Floor

The floor is composed of 1/4″ plate welded to embedded beams. The floor is covered with an 18″ concrete slab with a leak-chase system over all liner plate seams which are composed of 1/4″ butt welds.

2.2.4 Junctions

At the junctions where the cylinder intersects the dome and base slab, horizontal channels or angles are attached as anchors near the location of the maximum changes in meridional curvature.

2.3 Problem Areas

A section of the cylinder, after liner plate erection and the placing of concrete, is shown in Figure G-2. The majority of the loads imposed upon the liner (which will be discussed in Section 4) can be thought of as resulting from an inward radial movement and shortening of the concrete shell relative to the liner plate. The relative movement causes compressive membrane loads on the liner plate. If a panel of the liner plate should have an inward curvature

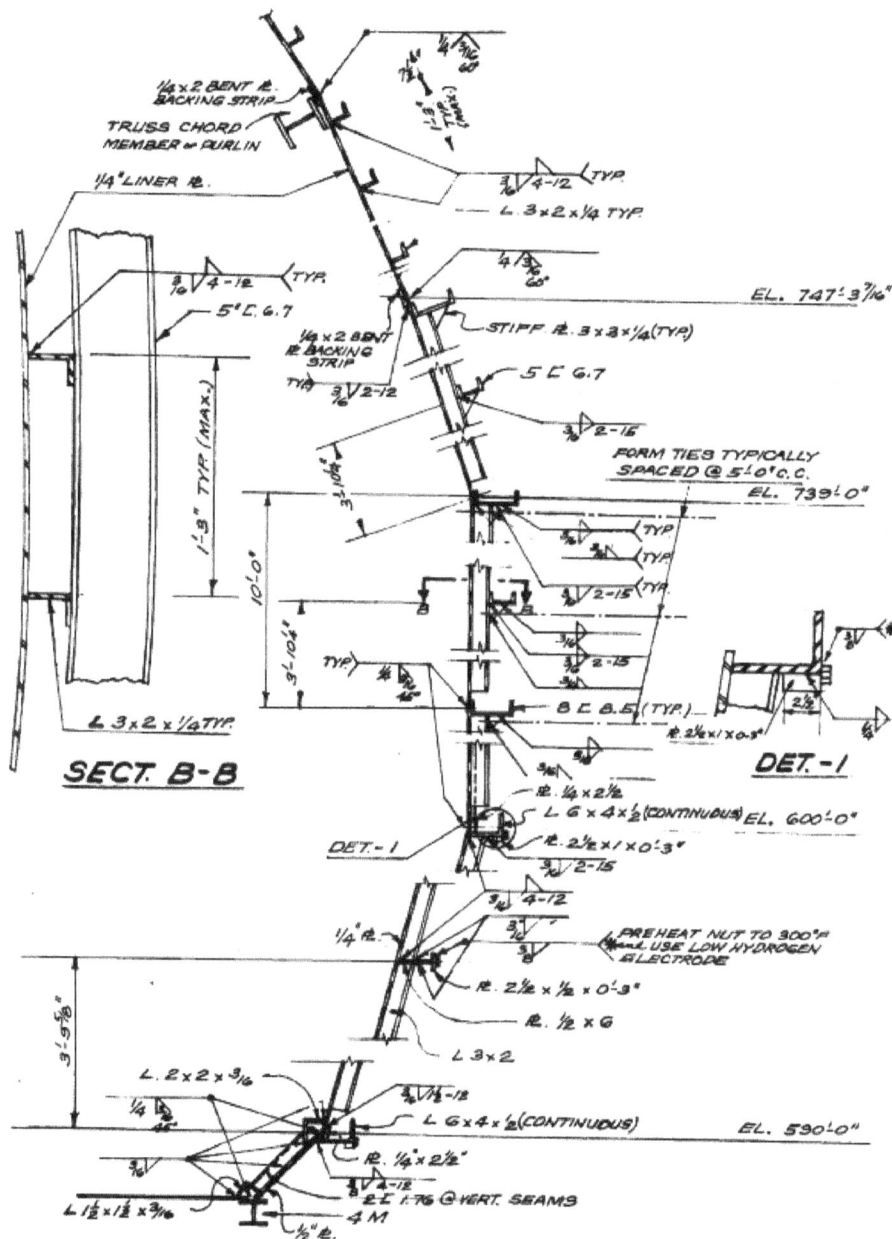

FIGURE G-1

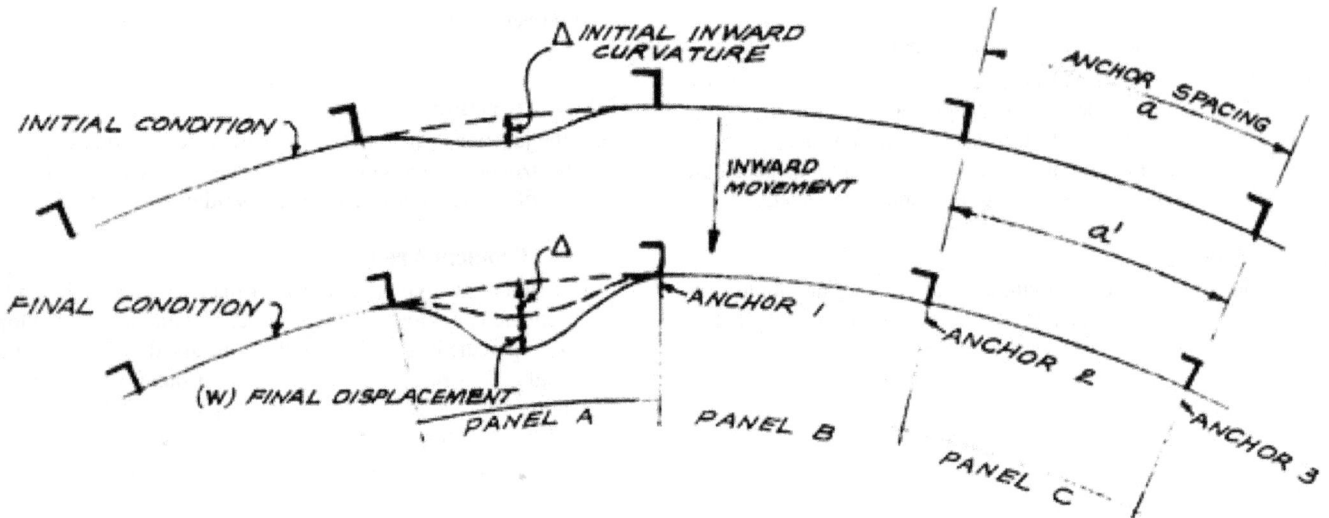

FIGURE G-2

resulting from fabrication and construction, such as illustrated in Figure G-2, then this panel will deform inward since it has low stiffness relative to a panel with outward curvature, and the anchorage system will be subjected primarily to a shear load. The analysis of this situation will be covered in Section 5. If the liner plate had and maintained a perfectly straight or curved geometrical shape, the liner anchors would never be required to withstand shear loads parallel to liner plate surface. The anchor will also be subjected to a moment, radial load and longitudinal strain; these items will also be discussed in Section 5.

The basic unit of the cylindrical liner plate is composed of stiffened panels whose length to width ratio is 8 to 1. Figure G-3 shows a possible deformation pattern for such a panel of liner plate. Due to that deformation pattern, the stiffening angles will be subjected to some torsion. The torsion will, however, be relatively minor since

it is a function of the slope of the curve shown as $\Delta X/\Delta Y$. Also, the analysis to be done for Figure G-2 is based on removing a strip of liner as shown in Figure G-3 which is at the most critical location and the solution involving the strip will not consider the resistance given by adjacent strips. This condition is much more severe on the anchorage system than any secondary torsional effects in the real system, and therefore, this problem will not be considered further.

The situation shown in Figure G-4, which is a longitudinal buckle of the cylinder, will be discussed further in Section 5. The discontinuities shown in Figure G-1 are in equilibrium when the liner plate is in compression and even local yielding of the concrete at the discontinuity cannot impair the integrity of the liner plate. If the concrete would be in tension at one of the discontinuity points, the liner plate integrity would still not be in jeopardy because, due to the amount of steel reinforcing used at discontinuities, cracking

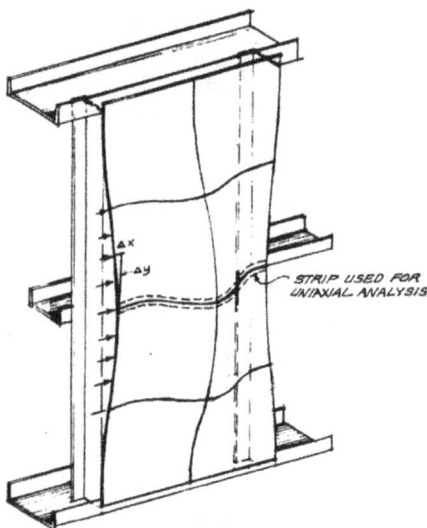

FIGURE G-3

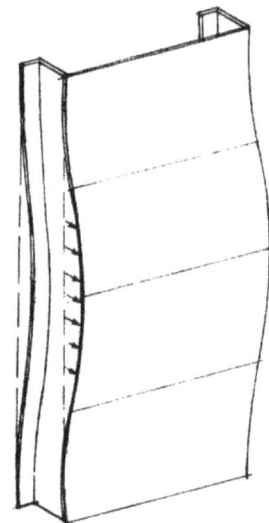

FIGURE G-4

is widely distributed and any pulling away of the liner plate would only lead to flexural liner strains of low magnitude. As shown in Figure G-1, there are anchors at these discontinuity points. The main function of the anchors is to stiffen the liner plate for erection and not to anchor the liner plate, even though they will serve in this capacity. Since the dome, which is shown in Figure G-1, is composed of flat stiffened support panels supported by trusses for erection purposes, the situation is not nearly as severe as that shown in Figure G-3, since all panels should have less relative variation in curvature than that of the cylinder, and therefore the analysis given for Figure G-2 will also cover the dome.

3. FACTORS AFFECTING THE LINER PLATE AND ANCHORAGE SYSTEM

3.1 Initial Inward Curvature of the Liner Plate

The fact that the liner plate will at some locations have initial inward curvature adjacent to panels that are flat or have outward curvature, is one of the most important items to consider when designing a liner plate anchor. The inward curvature in one panel enables the liner plate to relieve membrane loads leading to an unbalanced load condition at the anchor point and inducing loads into the anchor system. The larger the initial inward displacement, the greater is the ability of the panel to displace inward leading to greater loads imposed upon the anchor system. Due to this fact, the amount of inward curvature is limited to not more than 1/8″ during fabrication and erection of the liner plate. Inward curvature of 1/8″ is very likely to occur in some liner plate panels since the anchors are relatively close together and the plate is almost flat between the anchor points. As an example, if a plate with a radius of curvature of 58′ has anchors spaced at 15″, then the theoretical arc-to-chord distance will be only .04″, and minor external applied fabrication and erection loads will tend to snap the panel through leading to inward curvature.

3.2 Variation of Anchor Spacing

A variation in the anchor spacing will have the following effect on the liner plate anchorage system. If the anchors are further apart than the specified dimension, then a panel between anchors with inward curvature will have less stiffness than a panel with the specified dimension. This condition will lead to more shear load on the anchor. Due to the relatively Simple task of fabricating anchors to the specified spacing, any small variation in spacing is not considered to have any appreciable effect on the anchorage system.

3.3 Variation in Plate Thickness

Due to standard rolling tolerances, 1/4″ plate will have a thickness variation within the limits of +16% and −4%. The only effect of having a plate which is 4% under the theoretical thickness is that the plate would have a lower stiffness than the theoretical plate and there would be a slight increase in loading on the anchor. A plate which is thicker than the theoretical plate is advantageous as long as the excess thickness is constant throughout a large area, because a panel with inward curvature would be stiffer, and therefore, the anchor would be subjected to a decrease in load. In general, the smaller the ratio of anchor spacing to thickness the lower the loads in the anchorage system. It must be pointed out here that very low anchor spacing to plate thickness ratios tend to be unnecessary and quite uneconomical. The most undesirable condition is when the panel with inward curvature is thinner than the adjacent panels with outward curvature. The membrane loads imposed upon the panel with inward curvature are for this case greater, and the anchorage loads increase. In the analysis, a panel with outward curvature which is +16% over the nominal thickness will be considered adjacent to a plate with inward curvature of nominal thickness. The preceding condition is highly improbable and therefore, it is not necessary to consider the case of a plate which is −4% under the nominal thickness.

3.4 Variation in Liner Yield Stress

The liner plate, in general, is subjected to a strain input, as shown in Figure G-2, and the stress of the liner plate will increase linearly as a function of the concrete radial displacement only until the yield of the steel material is reached. After this point, strain will increase but the stress in the liner plate will remain essentially constant since the strains will exceed yield but will by no means approach ultimate. Therefore, with a fixed strain input, the upper bound for the state of stress in the liner plate will be the yield stress of the material. As an example, consider the strain in the liner plate to be .0015 in/in, then if the material yield stress is 30 ksi, the stress level will not exceed 30 ksi. If the material would have a yield stress of 60 ksi, then the stress level would not exceed 45 ksi since to reach 60 ksi, a strain of .002 in/in. would be required. Liner plates are just the opposite of normal structures; a low value of yield is advantageous since the low yield will limit the stress and force on the liner plate and the anchors. This phenomenon may be easily visualized by considering a liner plate made of rubber, since obviously this system will result in very small loads into the anchorage system.

3.5 Variation of Poisson's Ratio and Modulus of Elasticity of the Liner plate

Relative to other items which are subjected to variation, Poisson's ratio and Young's modulus of the liner plate will have very little significance on the performance of the liner plate and its anchorage system. As an example, approximately 52% during operation and 88% during the accident, of the load on the anchorage system is from the effects of thermal-gradients through the wall of the reactor building. Thermal strains are first calculated and then converted to thermal stress. In converting from strain to stress, the reciprocal of $(1-v^2)$ enters the expression and a variation of Poisson's ratio (v) has very little effect upon the calculated stress value. Variations in the modulus of elasticity of steel are not expected to have an appreciable effect upon the design of the liner plate, but the calculated value of thermal stress will vary with any change in modulus of elasticity and the amount of stiffness in a panel with inward curvature will also be affected by any variation in the modulus of elasticity of the liner plate. Due to the fact that only minor variations in the modulus of elasticity of steel are expected, this item will not be considered further.

3.6 Variation of Concrete Modulus of Elasticity and Anchor Stiffness

Since the anchorage system relies on both displacement and load-carrying capability, both of these items must be considered in evaluating the capability of the anchorage system. Displacement and force may be considered simultaneously by working in terms of energy which is really the same as being able to predict the point at which the system will be in equilibrium on its force vs.

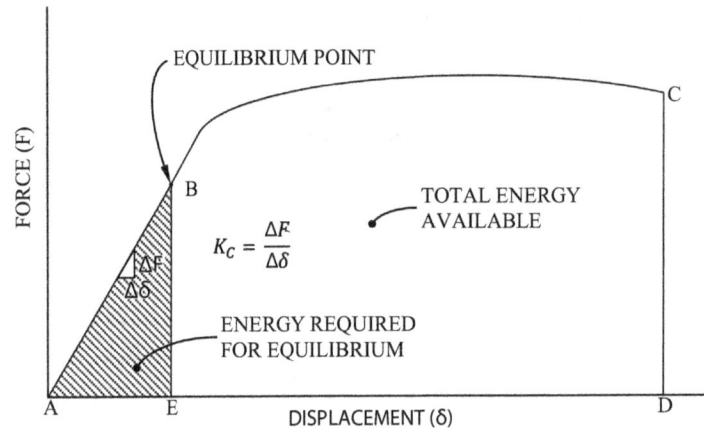

FIGURE G-5

displacement curve. Consider Figure G-5, which is similar to a typical load vs. displacement curve for an angle, channel, etc., embedded in concrete which resists shear forces parallel to the plate surface.

The total energy available in the anchorage system can be defined as the entire area (ABCD) under the curve. When subjected to a particular load, the point (B) on the curve at which the liner plate anchorage system is in equilibrium may be predicted. The area under the curve (ABE) up to this point may be considered the energy used up in obtaining equilibrium.

Dividing the total area (ABCD) by the equilibrium area (ABE) will result in a safety factor based on the ultimate energy capacity of the system. A variation in the modulus of elasticity of the concrete will affect the anchorage system as follows. When working in the linear range (AB) of the curve, the slope of the curve (which *is* also the spring constant of the system) will be a function of the modulus of elasticity of the concrete. The force in the anchorage system is equal to the displacement times the spring constant. The energy in the linear range of the curve is defined as one half (displacement) × (force), and substituting it may be noted that the energy equals one half (spring constant) × (displacement.) As the concrete modulus reduces, the spring constant reduces, but displacement increases. And since the energy involves the displacement squared, this results in more energy being required for equilibrium for low modulus of elasticity concrete.

3.7 Cracking or Crushing of Concrete in the Local Liner Plate Anchorage Zone

A liner plate anchorage system must have sufficient ductility and force carrying capability in order to accommodate the imposed loads. In general, embedded shear-carrying devices used in conjunction with concrete, exhibit the ability to deform when subjected to loads. When a shear-carrying device is loaded, the stress distribution exhibits a very high peak at the concrete surface, but this situation is quickly eliminated due to local yielding with the stresses further from the surface increasing in magnitude until an equilibrium condition is reached. Since concrete does not have the ability to greatly yield and develop high strains, crushing of the concrete takes place. The crushing of the concrete is necessary when testing a system to ultimate. When an anchorage system has sufficient safety factors, the local yielding of concrete will be of a very minor nature and certainly not an area of concern since it is

only the concrete's way of redistributing stress to obtain maximum load carrying capability.

3.8 The Tolerance of the Liner Plate and Anchorage System

Typical tolerances used in liner plate fabrication and erection are as follows:

1. The deviation in 15″ between liner stiffeners shall not exceed plus or minus 1/8″ when referenced to the theoretical surface.
2. A template 15′ in length shall not show deviations greater than plus or minus 3/4″. In the event that the template crosses a weld seam, the previous tolerance may be increased to plus or minus 1″.
3. In the longitudinal direction of the cylinder, a straight edge shall not show deviations greater than plus or minus 1/2″.
4. The slope of the cylindrical wall shall not exceed 1/240 within 10′ panels.
5. The misalignment of seams shall not exceed 10% of the plate thickness but for 1/4″ plate, the misalignment shall not exceed 1/16″.
6. The location of liner plate anchors relative to each other shall not deviate from the theoretical dimension by more than plus or minus 1/4″.

4. LOADS TO BE CONSIDERED IN THE LINER PLATE DESIGN AND THE STRAINS RESULTING FROM THESE LOADS

4.1 Individual Loads and Corresponding Liner Strain

4.1.1 Creep and Shrinkage

As the concrete dries out, it will tend to shrink, which, in general, will result in compressive strains in the liner plate. The expected value of shrinkage strain before pre-stressing is approximately $-100\ \mu$ in/in. The total shrinkage after pre-stressing will be combined with pre-stressing and given in a later section. No appreciable concrete creep is expected before pre-stressing and concrete creep will be combined directly with pre-stress in a later section.

4.1.2 Pre-Stress Combined with Concrete Creep and Shrinkage

The most critical situation is near end of life when the concrete has maximum creep and shrinkage. At this time, the dome liner will experience strains in the hoop and meridional direction of approximately −200 μ in/in. At the mid-height of the cylinder, the hoop strains will be approximately −334 μ in/in., and the meridional strain will be −69 μ in/in. The preceding low value is due to the fact that the pre-stressing force in the hoop direction is much greater than that of the meridional direction.

4.1.3 Deadload

The strain due to dead load at the mid-height of the cylinder is approximately −30 μ in/in in the meridional direction. The peak strain is in the meridional direction at the base of the cylinder and is approximately −45 μ in/in.

4.1.4 Earthquakes, Wind, Tornadoes, and Hydrostatic Loads

No appreciable hydrostatic loads are expected on the structure. In structures such as a reactor building housing Pressurized Water Reactors, the effects of high winds and tornadoes are negligible compared to the effects of earthquakes due to the large mass and stiffness of the structure. The strains from the maximum hypothetical earthquake are expected to vary from approximately 0 to ±76 μ in/in from the top of the structure to the base in the meridional direction, whereas the strain in the hoop direction should vary from 0 to ± 23 μ in/in.

4.1.5 Vacuum Loads

The design vacuum pressure is 3 psi. This pressure will generate a membrane hoop strain of −8.5 μ in/in and a membrane longitudinal strain of −3.0 μ in/in. A flexural strain of 180 μ in/in. in the liner plate will also exist between the anchors.

4.1.6 Pressure

The strain expected under accident conditions in the hoop and meridional direction in the dome should be approximately +100 μ in/in. The strains at the mid-height of the cylinder should be approximately ±50 μ in/in in the longitudinal direction and +147 μ in/in in the hoop direction.

4.1.7 Temperature

During the operating condition, the thermal strains in the cylinder and dome are expected to be approximately −325 μ in/in in both the hoop and meridional directions. This is based on a thermal gradient of 100°F on the inside face and 0°F on the outside face.

During the accident condition, the thermal strains are expected to be approximately −1500 μ in/in in the cylinder and the dome in both the hoop and meridional directions. These strains are based on the thermal gradient going from 280°F on the inside face to 0°F on the outside face with a very rapid drop-off in temperature within the first few inches of the inside face.

4.2 Load Combinations

The load combinations are shown in detail in the Detailed Analysis section. The normal operating condition is analyzed together with an accident in conjunction with the maximum hypothetical earthquake.

A protracted shutdown and a cold-weather start-up may yield conditions slightly more severe on the liner plate than the operating condition, but these conditions will not approach the accident condition with respect to harmful effects on the liner plate.

5. DETAILED ANALYSIS

5.1 General—Tangential Shear Load and Displacement of the Liner Plate Anchors

All of the major strains imposed on the liner plate such as concrete shrinkage and creep, pre-stressing and thermal effects may be considered as an inward movement of the shell as shown in Figure G-2. As the shell moves inward, all the panels will be subjected to membrane compression. But since Panel A has initial inward curvature, it will displace an additional amount (W) inward due to flexure. Since Panel A is not as stiff as the other panels, a tangential movement will occur, thus loading the anchor. As the tangential movement occurs at Anchor 1, the load in Panel B will drop off until Anchor 1 is in equilibrium. If only Panel A and Panel B are considered, then at equilibrium of Anchor 1, Anchor 2 will not be in equilibrium. To bring Anchor 2 in equilibrium, additional displacement will occur at Anchor 1. The effects of all anchors must be considered when analyzing Anchor 1.

5.1.1 Effects of All Anchors and Panels on Anchor 1 and Panel A

First consider the equilibrium of Anchor 1 after being subjected to the loads from Panel B. Figure G-6 shows the model of Anchor 1.

Description of Terms:

K_c': Spring constant 1st Anchor with concrete.
K_c: Spring constant of other Anchors with concrete.
K_{BPL}: Spring constant of Panel A or bent plate
K_{RPL}: Imaginary spring constant simulating load drop-off or relaxation of Panel B with Anchor 2 held fixed.
δ_{11}: Displacement of Anchor 1 due to unbalance at Anchor 1
N: Initial membrane force in Panel B

For equilibrium of Anchor 1 then

$$\delta_{11}(K_C + K_{BPL} + K_{RPL}) = N \quad \text{and}$$

(1) $$\delta_{11} = \frac{N}{K_C + K_{BPL} + K_{RPL}} \quad \text{and the force in the Anchor is}$$

(2) $$F_{11} = \delta_{11} K_C$$

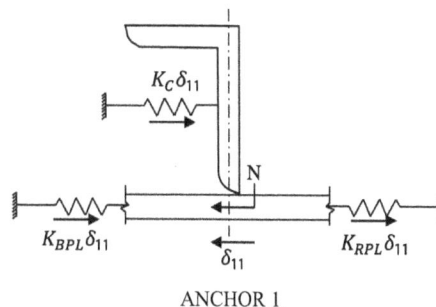

ANCHOR 1

FIGURE G-6

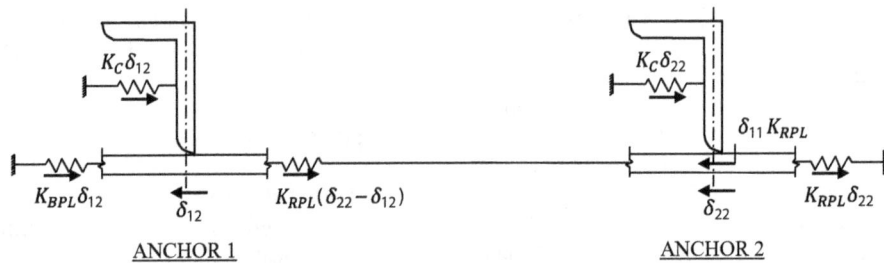

ANCHOR 1   ANCHOR 2

FIGURE G-7

Now since Anchor 1 has moved there is an unbalance at Anchor 2 equal to $\delta_{12} K_{RPL}$ and the following Figure G-7 will be used to illustrate how this unbalance affects Anchor 1.

In Figure G-7 δ_{22} is the displacement of Anchor 2 due to an unbalance at Anchor 2 and δ_{12} is the displacement of Anchor 1 due to an unbalance at Anchor 2. Figure G-8 shows Anchor 2 with a single spring used to represent Anchor 1.

For equilibrium of Anchor 2 then

$$\delta_{22}(K_C + K_{TI} + K_{RPL}) = \delta11 K_{RPL} \text{ and}$$

$$(3) \qquad \delta_{22} = \frac{\delta_{11} K_{RPL}}{K_C + K_{T1} + K_{RPL}}$$

Displacement at Anchor 1 due to a movement at Anchor 2

$$(4)\ \delta_{12} = \frac{\delta_{22}(K_{RPL})}{(K_C + K_{BPL} + K_{RPL})} \text{ Substitute (3) into (4) and simplify}$$

$$(5)\ \delta_{12} = \frac{\delta_{11}(K_{RPL})^2}{[(K_C + K_{BPL})(K_{RPL}) + (K_C + K_{RPL}) + (K_C + K_{BPL} + K_{RPL})]}$$

By combining Equations (1) and (5) the total displacement at Anchor 1 due to movement at Anchor 1 and 2 may be obtained

$$\delta_{11} + \delta_{22} = \frac{N}{(K_C + K_{BPL} + K_{RPL})}$$

$$(6)$$

$$\left[1 + \frac{(K_{RPL})^2}{(K_C + K_{BPL}) + (K_C + K_{RPL})(K_C + K_{BPL} + K_{RPL})}\right]$$

Now since Anchor 2 has moved, there is an unbalance at Anchor 3 equal to and Figure G-9 may be used to find how this unbalance affects Anchor 1.

Using the same approach which was used previously, the expression for the displacement at Anchor 1 due to movement of Anchors 1, 2, and 3 is found to be:

Since Term "A" in Equation 7 is always larger than 1.0 for the range of variables involved, it is conservative to assume that the equation for the displacement at Anchor 1, which considers the movement of an infinite number of anchors is as follows:

$$\delta = \sum_{n=1}^{n=\infty} \delta_{in} = \frac{N}{(K_C + K_{BPL} + K_{RPL})}$$

$$\left\{1 + \sum_{n=1}^{n=\infty}\left[\frac{K_{RPL}^2}{(K_C + K_{BPL})(K_{RPL}) + (K_C + K_{RPL})(K_C + K_{BPL} + K_{RPL})}\right]^n\right\}$$

$$(8)$$

A very convenient way to handle this problem and reduce it to a one anchor solution is as follows: Let N' be an imaginary load which simulates the effects of all anchors causing moment at Anchor 1

$$N' = N\left\{1 + \sum_{n=1}^{n=\infty}\left[\frac{(K_{RPL})^2}{(K_C + K_{BPL})(K_{RPL}) + (K_C + K_{RPL})(K_C + K_{BPL} + K_{RPL})}\right]^n\right\}$$

$$(9)$$

Then the single anchor solution and model are as shown in Figure G-10

$$(10) \qquad \delta = \frac{N'}{(K_C + K_{BPL} + K_{RPL})} \text{ Total displacement of Anchor 1 in the linear range}$$

$$(11) \qquad F_C = \delta K_C \qquad \text{Force in anchor}$$

$$(12) \qquad F_{BPL} = \delta K_{BPL} \qquad \text{Force in bent plate}$$

$$\delta_{11} + \delta_{12} + \delta_{13} = \frac{N}{(K_C + K_{BPL} + K_{RPL})}\left\{1 + \frac{K_{RPL}^2}{(K_C + K_{BPL})(K_{RPL}) + (K_C + K_{RPL})(K_C + K_{BPL} + K_{RPL})}\left[1 + \right.\right.$$

$$\left.\left.\frac{K_{RPL}^2}{\{(K_C + K_{BPL})(K_{BPL}) + (K_C + K_{RPL})(K_C + K_{BPL} + K_{RPL})\}\left\{\frac{1}{K_C + K_{BPL} + K_{RPL}}\left[(K_C + K_{RPL}) + \frac{K_{RPL}[(K_C + K_{BPL})(K_{RPL}) + (K_C) + (K_C + K_{BPL} + K_{RPL})]}{[(K_C + K_{BPL})(K_{RPL}) + (K_C + K_{RPL})(K_C + K_{BPL} + K_{RPL})]}\right]\right\}}\right]\right\}$$

TERM–"A"

$$(7)$$

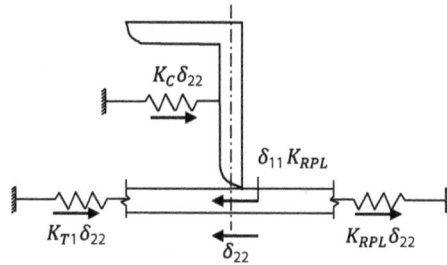

FIGURE G-8

In Figure 8

$$K_{T1} = \frac{(K_C + K_{BPL})(K_{RPL})}{(K_C + K_{BPL} + K_{RPL})}$$

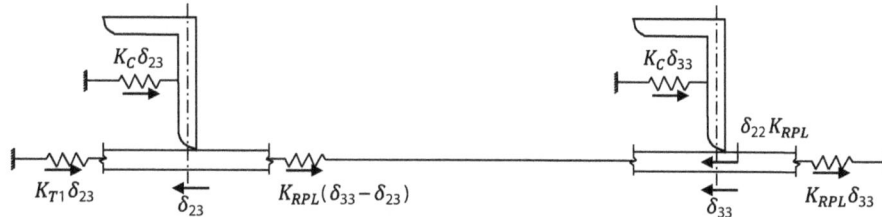

FIGURE G-9

5.1.2 General Description of Necessary Information and Analysis Techniques

In order to evaluate the load and displacement of the anchor shown in Figure G-10, the following information is needed:

(1) A load vs. displacement curve of the anchor embedded in concrete as shown in Figure G-11

(2) A load vs. displacement curve of the bent plate as shown in Figure G-12

After obtaining the information shown in Figures G-11 and G-12 by testing or any other reliable method, the anchor may be analyzed.

Figure G-12 shows all information needed for the solution.

When the required anchor displacement is low enough so that the load vs. displacement curves for the anchor and the bent plate are in the linear range then the following elastic solution may be used:

Elastic Solution

$$\delta = \frac{N'}{(K_C + K_{BPL} + K_{RPL})} \quad \text{and } F_C = \delta K_C \quad F_{BPL} = \delta K_{BPL}$$

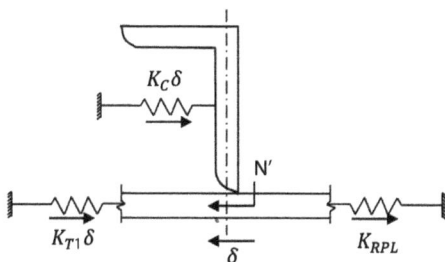

FIGURE G-10

When the required anchor displacement is great enough so that the displacement is in the nonlinear range of one of the curves, then the following Plastic Solution must be used.

Plastic Solution

For ease in visualizing this solution an example problem will be used.

Consider the following anchor (as shown in Figure G-14) with only the concrete resisting the imposed load. Since the load N' is self-limiting, K_{RPL} will simulate the load decrease as displacement occurs. Also included in Figure G-14 is the information applicable to the anchor and relaxation characteristics.

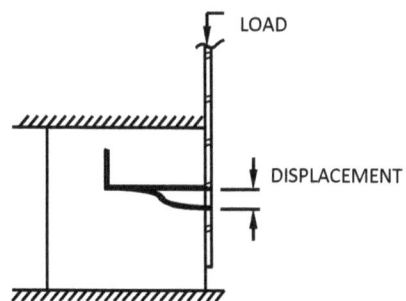

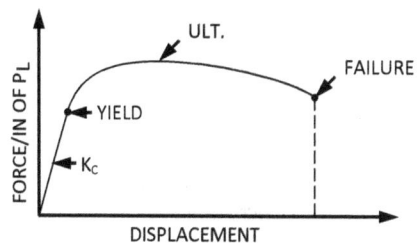

FIGURE G-11

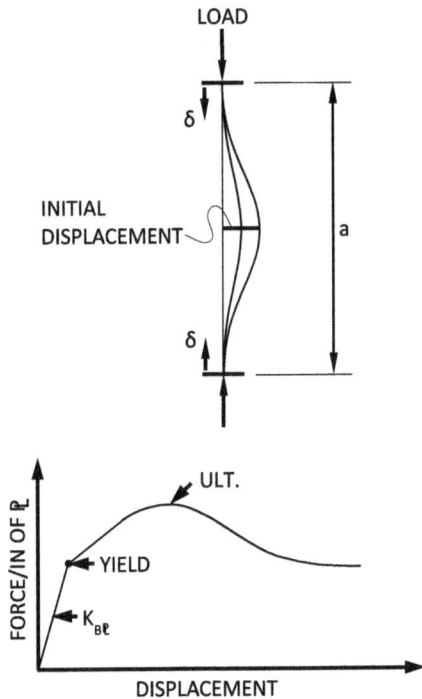

FIGURE G-12

Let $N' = 28$ K/in and use an elastic solution to see if the anchor is in the plastic range:

$\delta \dfrac{28}{(200+500)} = 0.4''$. Since δ is greater than $0.25''$, the solution must consider the plastic range. When a displacement of $.025''$ is reached, the unbalanced load (F) is as follows:

$$F = N' - \delta_e(K_C + K_{RPL}) = 28 - .025(200 + 500) = 10.50 \text{ K/in}.$$

After a displacement of $.025''$ is reached, then only plate relaxation is available to reach equilibrium and the additional displacement (δ) is:

$$\delta' = \frac{F}{K_{RPL}} = \frac{10.50}{500} = 0.21'' \text{ The total displacement is } \delta = \delta e + \delta' = .025 + .021 = .046''$$

The energy used up in obtaining equilibrium in the anchor is:

$$E = \frac{1}{2}(5)(.025) + (5.0).46 - .025) = .1675 \text{ K/in}$$

This energy is shown as the shaded area under Anchor Curve in Figure G-14. The total energy in the anchor is defined as the total area under the Anchor Curve. The total energy is:

$$E_T = \frac{1}{2}(5)(0.25) + (5.0)(.150 - .025) = .6875 \text{ K/in}.$$

The safety factor may be defined as:

$$S.F. = \frac{E_T}{E} = \frac{.6875}{.1675} = 4.1$$

5.1.3 Modification of Test Results for Use in the Analysis

The tests performed to determine the characteristics of the bent plate used a material which had a yield strength of 43 ksi. Since data will be needed for yield strengths both higher and lower than 43 ksi, the original data will be modified as shown below in Figure G-15. The shifting technique used should be fully valid since with an initial eccentricity of $1/8''$ the plate is primarily in bending and the curve is never shifted high enough to approach the buckling load. See Appendix A for test results.

Figures G-16 and G-17 show the data which will be used for the properties of the bent plate.

The test performed to determine the anchor characteristics had a concrete modulus of elasticity of 2.67×10^6 psi at the time of testing. Since a higher modulus of elasticity is expected in the real structures, the anchor load vs. displacement curves will be modified as shown in Figure G-18. It is assumed that the anchor spring constant changes linearly as a function of the change in modulus of elasticity. For test results see Appendix B.

Figures G-19 through G-21 show the data which will be used for the properties of the anchors. It may be noted in Appendix B that the least value of total energy was .541 K/in/in. and this occurred with a maximum force of 5.97 K/in. To be conservative, the total energy will be assumed to be .541 K/in/in. and the maximum force will be limited to 5.00 K/in.

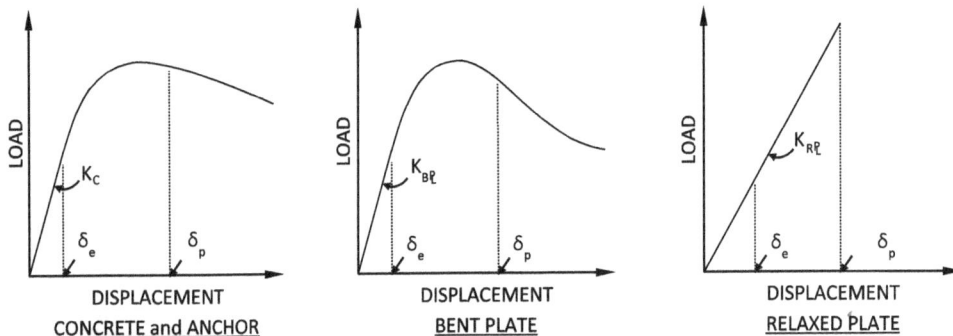

FIGURE G-13

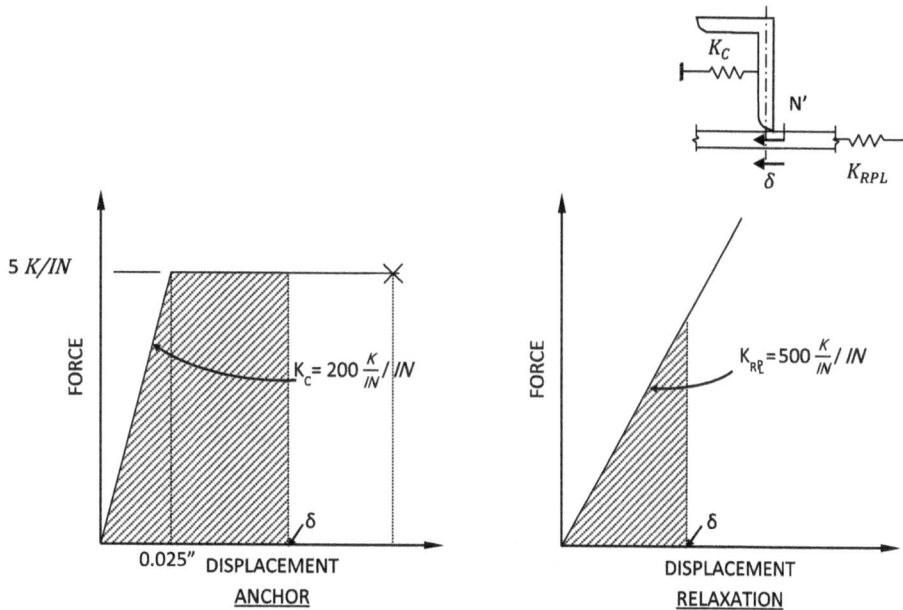

FIGURE G-14

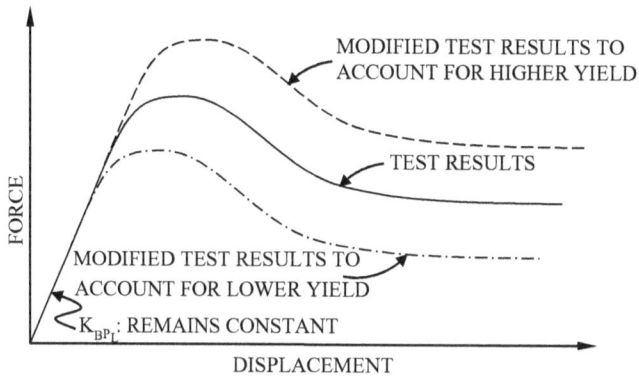

FIGURE G-15

5.1.4 Effects of Internal and External Pressure on the Liner Plate Anchors

Internal pressure results due to a loss of coolant accident. This pressure will tend to reduce the shear load on the anchorage system as shown in Figure G-22. Also, the pressure will provide restraint against any minimal pullout forces exerted on the anchorage system during the accident.

External pressure is exerted from a vacuum load which results from the difference in pressure on the inside and outside of the reactor building. The vacuum pressure acts in the opposite direction of that shown in Figure G-22.

The following analysis will illustrate how the pressure may be converted to an equivalent axial load. This axial load will be very convenient when analyzing individual anchors. The applicable

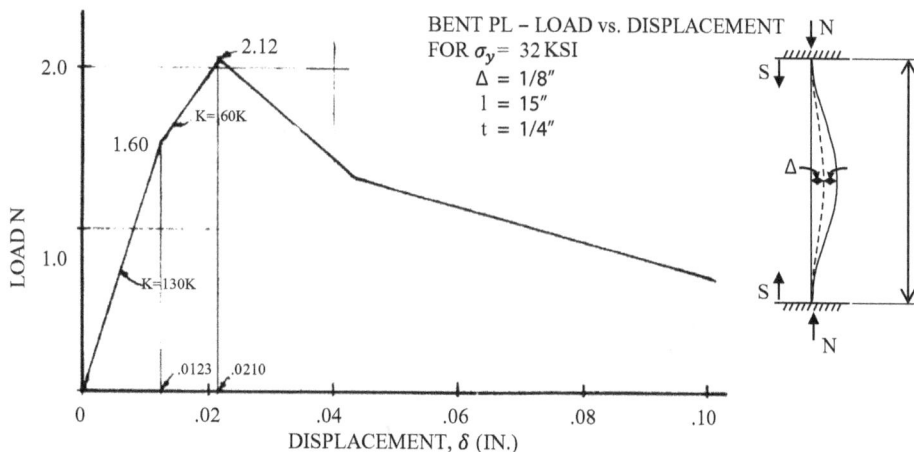

FIGURE G-16

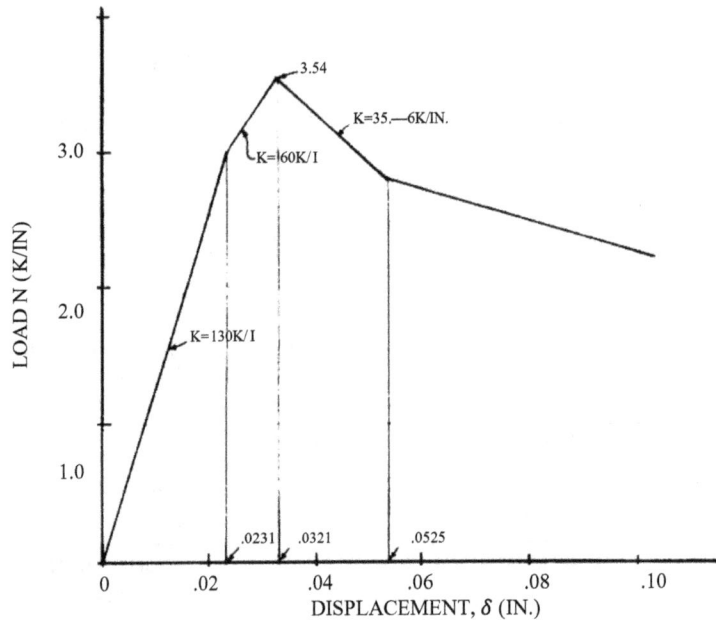

FIGURE G-17

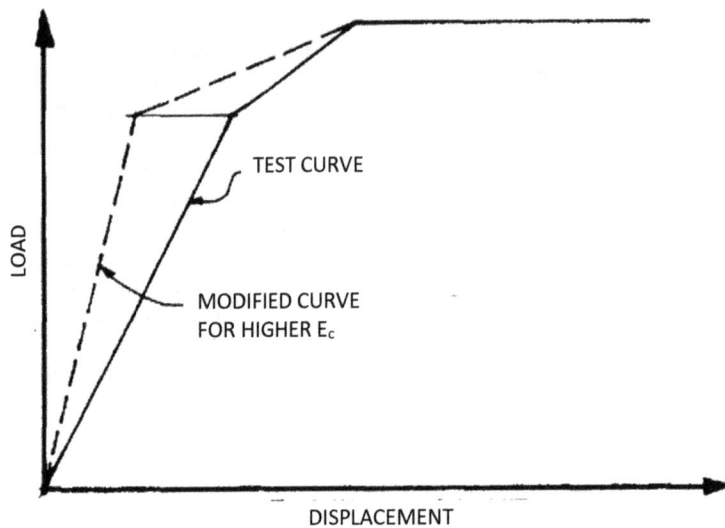

FIGURE G-18

differential equation for loading conditions shown in Figure G-22 is as follows:

$$(13) \qquad \frac{EId^4W}{dx^4} + \frac{Nd^2W}{dx^2} + \frac{Nd^2\Delta}{dx^2} = -q$$

Using the following relationships and Figure G-23

$$W = W_m \sin^2\frac{\pi}{a}x \xi \Delta = \Delta_m \sin^2\frac{\pi}{a}x$$

An appropriate solution for this problem is as follows:

$$(14) \qquad W_m = \frac{\Delta_m N - \dfrac{qa^2}{2\pi^2}}{\left(\dfrac{4\pi^2 EI}{a^2} - N\right)}$$

The axial load equivalent to the pressure effect may be found by the following:

$$\Delta_m N'' = \frac{qa^2}{2\pi^2}$$

$$(15)$$

$$N'' = \frac{qa^2}{2\pi^2 \Delta_m}$$

5.1.5 Detailed Analysis of Anchorage System

As a review, a detailed outline of the method of analysis will be given below:

1. Combine the strains from all loads being considered and determine the two-dimensional state of strain. Using the largest of the strains, determine the uniaxial yield stress that must exist if the strains are to be converted to stress by use

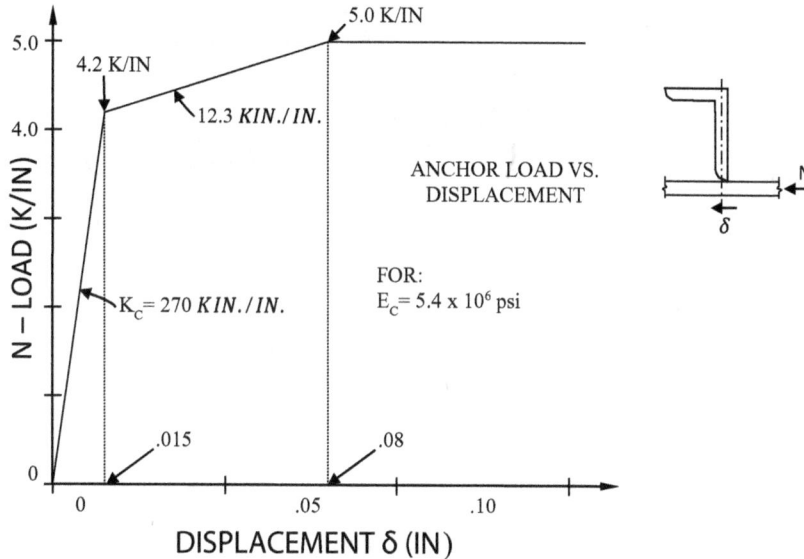

FIGURE G-19

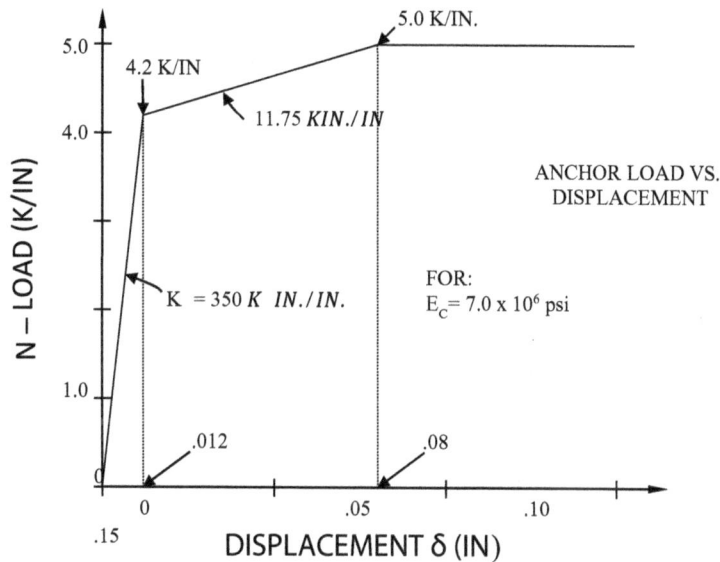

FIGURE G-20

of Hooke's Law. Convert strains to stress and determine the load per inch. This portion is highly conservative due to the fact that a strip of the shell is being analyzed, but the imposed load comes from a biaxial state of stress. Using a uniaxial yield to determine the bent plate characteristics is also conservative since it neglects the stiffening and increase in yield due to a biaxial state of stress.

2. Determine the load vs. displacement curves for the anchor and bent plate which are applicable to the problem based on the concrete modulus and steel yield strength.

3. Determine the spring constant for the plate relaxation based on $K_{RPL} = \dfrac{AE}{a}$.

4. Find the force N′ which simulates the effects of all other anchor movements use Equation 9.

5. If internal or external pressure is acting on the liner plate, then the pressure load will be converted to an equivalent axial load (N″) by use of Equation 15 and combined with N′.

6. Do an Elastic Solution of the anchor to determine if either the anchor or bent plate is over yield. If either or both is over yield, then do a Plastic Solution.

The following cases will be analyzed for their effects on the Anchorage System.

Case I: Most probable worst condition to which the anchor is subjected during operation with $E_C = 5.4 \times 10^6$ psi and $\sigma_y = 32$ ksi.

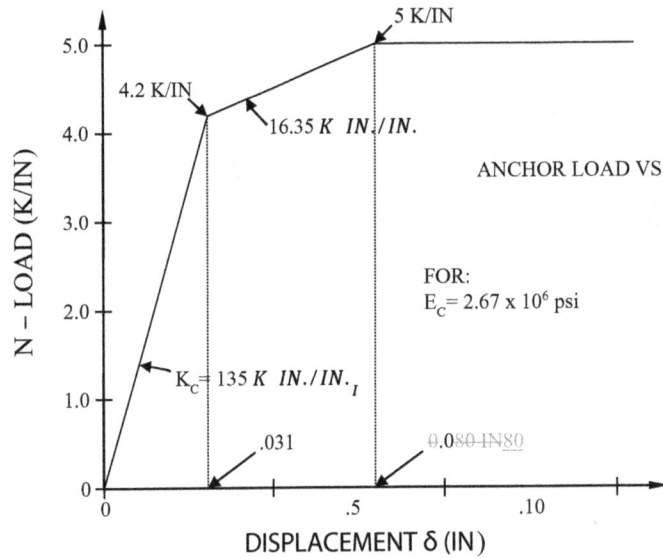

FIGURE G-21

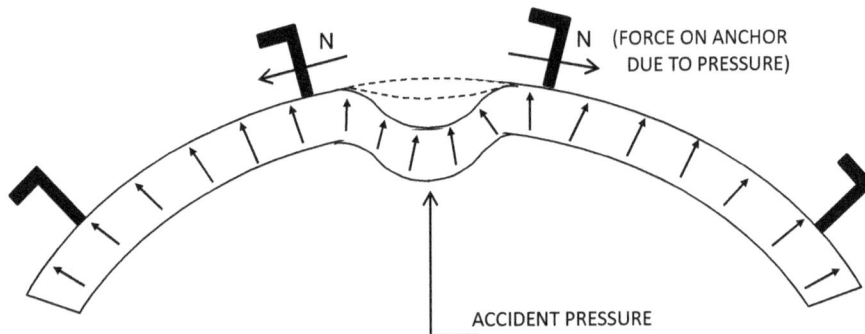

FIGURE G-22

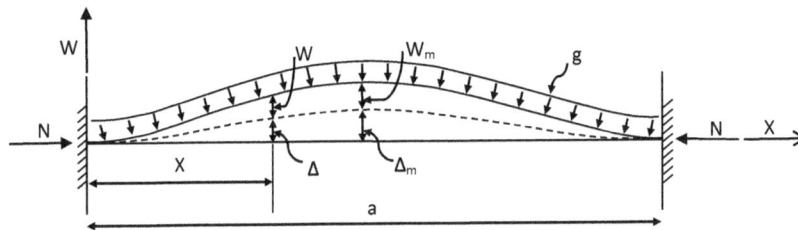

FIGURE G-23

Case II: Most probable worst condition to which the anchor is subject in the event of a loss of coolant accident with psi and ksi.

Case III: Same as Case II but $E_C = 7.0 \times 10^6$ psi

Case IV: Same as Case II but $E_C = 2.67 \times 10^6$ psi

Cases III and IV are being analyzed to show how the variation of concrete modulus affects the anchorage system.

Case I

<u>Operating Condition with Vacuum</u>

State of stress and strain in liner at mid-height

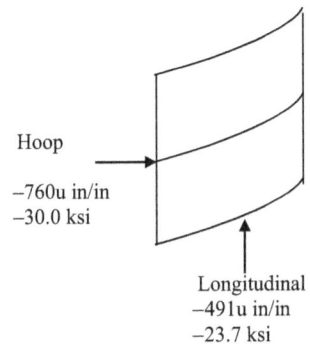

Hoop
−760u in/in
−30.0 ksi

Longitudinal
−491u in/in
−23.7 ksi

STRAIN TO STRESS CONVERSION

$$E = 30 \times 10^3 \text{ ksi} \qquad v = .3$$

$$\sigma_x = \frac{E}{(1-v^2)}(\varepsilon_x + v\varepsilon_y)$$

Summary of Strains

Source	Hoop	Longitudinal
1. Prestress w/ creep & shrinkage	−334u	−69u
2. Dead load	+10u	−30u
3. Earthquake max. hypo.	±11u	±33u
4. Thermal 100°F Inside 0°F Outside	−325u	−325u
5. Initial shrinkage before prestress	−100u	−100u
TOTAL	−760u	−491u

Miscellaneous Information:

1. Uniaxial yield strength of material is at least 32 ksi
2. Plates adjacent to the panel with initial curvature are 16% over standard thickness.
   Plate thickness:

$$h = \frac{1}{4}(1.16) = 0.29 \text{ in.}$$

Maximum membrane stress

$$\sigma_x = \frac{E}{(1-v^2)}(\varepsilon_x + v\varepsilon_y) = \frac{30000}{(1-0.30^2)}(760 + 0.3(491))10^{-6} =$$

$$\sigma_x = 30.0 \text{ k/in}^2 < \sigma_y = 32.0 \text{ K/in}^2$$

Maximum Membrane Force:

$$N = h(\sigma x) = 0.29(30.0) = 8.72 \text{ K/in}$$

Assumed concrete modulus of elasticity $E = 5.4 \times 10^3$ ksi with an anchor spring constant in the linear range of $K_c = 270$ K/in/in. (See Figure G-19).

The bent plate will have a spring constant of $K_{BPL} = 130$ K/in/in in the linear range (See Figure G-16).

The plate when relaxing will have a spring constant of:

$$K_{RPL} = \frac{AE}{a}(1.16) = \frac{(.25)(30\times10^3)(1.16)}{15} = 580 \text{ K/in/in}$$

Find the force to put on the first anchor which simulates the effects of all other anchors on the first anchor:

$$N' = N\left\{1 + \sum_{n=1}^{n=\infty}\left[\frac{K_{RPL}^2}{(K_C + K_{BPL})(K_{RPL}) + (K_C + K_{RPL})(K_C + K_{BPL} + K_{RPL})}\right]^n\right\}$$

Using $N = 8.72$ K/in $\quad K_{RPL} = 580$ K/in/in $\quad K_C = 270$ K/in/in $\quad K_{BPL} = 130$ K/in/in

$$N' = 8.72\left\{1 + \sum_{n=1}^{n=\infty}\left[\frac{580^2}{(400)(580) + (580)(980)}\right]^n\right\}$$

$$= 8.72\{1 + \sum_{n=1}^{n=\infty}(.316)^n\} = 12.76 \text{ k/in}$$

Find the additional axial load which simulates a 3 psi vacuum:

$$N'' = \frac{qa^2}{2\pi^2\Delta_m} = \frac{(.003)(225)}{(2)(9.85)(.125)} = .281 \text{ K/in}$$

$$N_T = N'' + N' = 12.76 + .28 = 13.04 \text{ K/in}$$

Anchor Problem

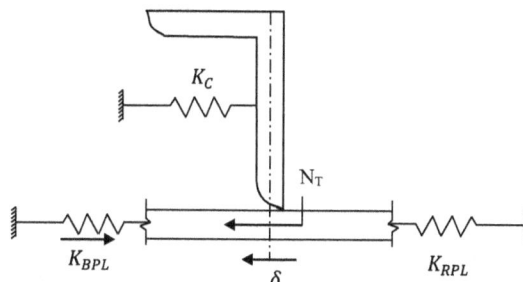

Use Figures G-16 and G-19 for the bent plate and anchor properties.

Elastic Solution

$$\delta(K_C + K_{BPL} + K_{RPL}) = N_T \qquad \delta = \frac{N_T}{(K_C + K_{BPL} + K_{RPL})} = \frac{13.04}{980} = .0133''$$

Since the yield of the bent plate is exceeded, a Plastic Solution must be used:

$$\underset{\text{Anchor}}{270(\delta)} + \underset{\text{Bent Plate}}{1.6 + 60(\delta - .0123)} + \underset{\text{Relaxation}}{580(\delta)} = \underset{\text{Load}}{13.04}$$

Solving the above equation $\delta = .0134''$

Check equilibrium:

Anchor:	$F_C = 270(.0134)$	= 3.62 k/in
Bent Plate:	$F_{BPL} = 1.6 + 60(.0134 - .0123)$	= 1.67 k/in
Relaxation:	$F_{RPL} = 580(.0134)$	= 7.78 k/in

Σ Forces = 13.07 ≈ 13.04 within slide rule accuracy.

The energy used up in obtaining equilibrium in the anchor is:

$$E = \frac{1}{2}(3.62)(.0134) = .0242 \text{ K-in/in}$$

The safety factor is:

$$S.F. = \frac{E_T}{E} = \frac{.541}{.0242} = 22.4$$

Case II
Accident Condition
State of stress and strain in liner at mid-height

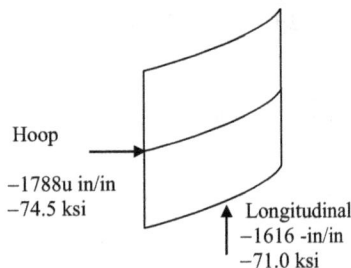

Hoop

$-1788u$ in/in
-74.5 ksi

Longitudinal
-1616 -in/in
-71.0 ksi

Miscellaneous Information:

1. In order to get the above stresses, then a uniaxial yield stress of $\sigma_y = (1788)(30) = 53700$ psi must exist.
2. Plates adjacent to the panel with initial curvature are 16% over thickness
3. Internal pressure is 55 psi
   Maximum membrane force $N = (74.5)(1/4)(1.16) = 21.6$ K/in.

Summary of Strains

Source	Hoop	Longitudinal
1. Prestress w/ creep and shrinkage	$-334u$	$-69u$
2. Dead load	$+10u$	$-30u$
3. Earthquake max. hypo.	$\pm11u$	$\pm33u$
4. Thermal 100°F Inside 0°F Outside	$-1500u$	$-1500u$
5. Initial shrinkage before prestress	$-100u$	$-100u$
6. Accident pressure	$+147u$	$+50$
TOTAL	$-1788u$	$-1616u$

Assumed concrete modulus of elasticity $E = 5.4 \times 10^3$ ksi with an anchor spring constant in the linear range of $K_C = 270$ K/in/in. (See Figure G-19)

The bent plate will have a spring constant of $K_{BPL} = 130$ K/in/in. in the linear range (See Figure G-17).

The plate when relaxing will have a spring constant of $K_{RPL} = \frac{AE}{2}(1.16) = 580$ K/in/in.

Find the force to put on the first anchor which simulates the effects of all other anchors on the first anchor by using equation 9 together with: $N = 21.6$ K/in/in. $K_{RPL} = 580$ K/in/in. $K_C = 270$ K/in/in. $K_{BPL} = 130$ K/in/in.

$$N' = 21.6(1.452) = 31.4 \text{ K/in.}$$

Find the force which must be subtracted from N' to simulate the effect of a 55 psi internal pressure:

$$N'' = \frac{qa^2}{2\pi^2\Delta_m} = \frac{(.055)(225)}{(2)(9.85)(.125)} = 5.03 \text{ K/in.}$$

$$N_T = 31.4 - 5.03 = 26.37 \text{ K/in.}$$

Use Figures G-17 and G-19 for the bent plate and anchor properties – try an Elastic Solution

$$\delta = \frac{N_T}{(K_C + K_{BPL} + K_{RPL})} = \frac{26.37}{980} = .0270''$$

Since both the yield of the bent plate and anchor are exceeded, a Plastic Solution must be used:

Plastic Solution
Assuming the displacement is less than .0321″ (See Figure G-17), then the following equilibrium equation may be written:

$$\overbrace{4.2 + 12.3(\delta - .015)}^{\text{Anchor}} + \overbrace{3.0 + 60(\delta - .0231)}^{\text{Bent Plate}} + \overbrace{580(\delta)}^{\text{Relaxation}} = \overbrace{26.37}^{\text{Load}}$$

Solving the above equation $\delta = .0318''$

Check equilibrium:

Anchor:	$F_C = 4.2 + 12.3(.0318 - .0150)$	$= 4.41$ k/in
Bent Plate:	$F_{BPL} = 3.0 + 6.0(.0318 - .0231)$	$= 3.52$ k/in
Relaxation:	$F_{RPL} = 580(.0318)$	$= 18.50$ k/in

Σ Forces $= 26.43 \approx 26.37$ within slide rule accuracy.

The energy used up in obtaining equilibrium in the anchor is:

$$E = \frac{1}{2}(4.2)(.015) + 4.2(.0318 - .015)/2 = .1039 \text{K-in/in}$$

The safety factor is:

$$S.F. = \frac{E_T}{E} = \frac{.541}{.1039} = 5.2$$

Case III
Accident Condition:
All data is the same as Case II except that the concrete modulus is $E_C = 7 \times 10^6$ psi. The anchor spring constant is $K_C = 350$ K/in/in. (See Figure G-20).

Find the force to put on the first anchor which simulates the effects of all other anchors on the first anchor by using equation 9 together with: $N = 21.6$ K/in, $K_{RPL} = 580$ K/in/in, $K_C = 350$ K/in/in, $K_{BPL} = 130$ K/in/in.

$$N' = 21.6(1.363) = 29.5 \text{ K/in}$$

The force which must be subtracted from N' to simulate the effect of a 55 psi internal pressure: N″ = 5.03 K/in.

$$N_T = 29.5 - 5.03 = 24.47 \text{ K/in.}$$

Based on Case II it is known that a Plastic Solution is necessary and the displacement is less than .0321″.

Plastic Solution

$$\overbrace{4.2 + 11.75(\delta - .012)}^{\text{Anchor}} + \overbrace{3.0 + 60(\delta - .0231)}^{\text{Bent Plate}} + \overbrace{580(\delta)}^{\text{Relaxation}} = \overbrace{24.47}^{\text{Load}}$$

Solving the above equation $\delta = .0288''$

Check equilibrium:

Anchor: $F_C = 4.2 + 11.75 (.0288 - .0120)$ = 4.39 k/in
Bent Plate: $F_{BPL} = 3.0 + 60(.0288 - .0231)$ = 3.42 k/in
Relaxation: $F_{RPL} = 580(.0288)$ = 16.70 k/in

$$\Sigma \text{ Forces} = 24.51 \approx 24.47$$

The energy used up in obtaining equilibrium in the anchor is:

$$E = \frac{1}{2}(4.2)(.012) + 4.2(.288 - .012) + (4.39 - 4.2)$$

$$(.0288 - .012)/2 = .0974 \text{ K-in/in}$$

The safety factor is:

$$S.F. = \frac{E_T}{E} = \frac{.541}{.0974} = 5.56$$

Case IV
Accident Condition:
All data is the same as Case II except the concrete modulus is $E_c = 2.67 \times 10^6$ psi. The anchor spring constant is $K_c = 135$ K/in/in. (See Figure G-21).
Find the force to put on the first anchor which simulates the effects of all other anchors on the first anchor by using equation 9 together with: $N = 21.6$ K/in/in, $K_{RPL} = 580$ K/in/in, $K_c = 135$ K/in/in, and $K_{BPL} = 130$ K/in/in.

$$N' = 21.6(1.795) = 38.8 \text{ K/in.}$$

The force which must be subtracted from N' to simulate the effect of a 55 psi internal pressure: $N'' = 5.03$ K/in.

$$N_T = 38.8 - 5.03 = 33.77 \text{ K/in}$$

Try an Elastic Solution to find the appropriate value of $\delta = \frac{33.77}{(135 + 130 + 580)} = .04''$

Plastic Solution

$$\underbrace{4.2 + 16.35(\delta - .031)}_{\text{Anchor}} + \underbrace{3.54 - 35.6(\delta - .0321)}_{\text{Bent Plate}} + \underbrace{580(\delta)}_{\text{Relaxation}} = \underbrace{33.77}_{\text{Load}}$$

Solving the above equation $\delta = .0455''$

Check equilibrium:

Anchor: $F_C = 4.2 + 16.35(.0455 - .031)$ = 4.44 k/in
Bent Plate: $F_{BPL} = 3.54 - 35.6(.0455 - .0321)$ = 3.06 k/in
Relaxation: $F_{RPL} = 580(.0455)$ = 26.40 k/in

$$\Sigma \text{ Forces} = 33.90 \approx 33.77$$

The energy used up in obtaining equilibrium in the anchor is:

$$E = \frac{1}{2}(4.2)(.031) + 4.2(.0455 - .031) + (4.44 - 4.20)$$

$$(.0455 - .031)/2 = .128 \text{ K-in/in}$$

The safety factor is:

$$S.F. = \frac{E_T}{E} = \frac{.541}{.128} = 4.23$$

The results of Cases I to IV are summarized below:

Condition	Case I Operating	Case II Accident	Case III Accident	Case IV Accident
Concrete Modulus (psi)	5.4×10^6	5.4×10^6	7.0×10^6	2.67×10^6
Anchor Spring Const. K/in/in.	270	270	350	135
Anchor Disp. (in.)	.0134	.0318	.0288	.0455
Anchor Force K/in.	3.62	4.41	4.39	4.44
Safety Factor	22.4	5.20	5.56	4.23

[G-2] SUMMARY CONCRETE CONTAINMENT VESSEL LINER PLATE ANCHORS AND STEEL EMBEDMENT TEST RESULTS, T.E. JOHNSON, P.C. CHANG-LO AND B.P. PFEIFFER, BECHTEL POWER CORPORATION, SMIRT 4, 1977 [J5/9].

This paper summarizes test data on shear load and deformation capabilities for liner plate line anchors in reinforced and prestressed concrete containments and structural steel embedments in reinforced concrete.

Static load versus displacement test data are necessary to assure that the design is adequate for the maximum loads. In addition, the ASME Section III, Division 2 requires that the liner design consider fatigue.

The line anchor test program had the following objectives:

(1) Determine load versus displacement data for a variety of line anchors utilizing structural tees and small beams with different weld configurations.
(2) Test line anchors under cyclic loads resulting from annual thermal transients.

Steel embedments in the containment and other structures may be subjected to pipe rupture effects which result in' shear. Since these loads are of a transient type, it Is necessary to obtain the static load-displacement relationship so that the energy absorption capacities of the steel embedments can be determined.

The steel embedment test program had the following objectives;

(1) Determine load versus displacement relationship for square and rectangular steel embedments with several cross-sectional areas.
(2) Det ermine the maximum capacities of the steel embedments with varying width-to-thickness ratios and constant cross-sectional area.

[G-3] INCREASE IN STEEL LINER STRAINS DUE TO CONCRETE CRACKING—A REEXAMINATION OF THE TENSION LINER STRAIN ALLOWABLES, O. JOVALL AND P. ANDERSON, SCANSCOT TECHNOLOGY AB, JUNE, 2008.

Abstract

In Article CC-3710, liner strain allowables for Service and Factored loads are specified. Calculated liner strains, when specified loads are applied, shall not exceed the values given in Table CC-3720-1. The design calculations of the liner strains do not normally explicitly include the effect of concentration of plastic liner strains due to concrete cracking. The aim of this report is to investigate the level of tensile plastic strain reached locally due to concrete cracking, if the strain level in the liner adjacent to the cracked area has reached the maximum allowable value specified in the ASME Code (Table CC-3730-1), $\varepsilon_{st} = 0.003$. Additionally, the combined effect of concrete cracking, the friction between the liner and the concrete wall, and discontinuities such as local thinning of the liner, is discussed. The calculated maximum plastic strains at a concrete crack is then compared to the estimated strain level when liner tearing may occur. This to investigate the safety margin regarding the localized plastic strains in the liner then occurring, to the strain level when liner tearing may occur.

From this study, the following conclusions can be drawn:

- Cracking of the concrete containment wall will give rise to localization of the tensile liner strain.
- The friction between the liner and the concrete wall will concentrate the plastic strains to through-wall concrete cracks, giving a further increase of the localized strain level.
- Due to the influence of friction the liner cannot be stretched beyond the hardening part. When the characteristic steel curve is flattened out after the hardening part, the remaining displacement will be concentrated to the crack location.
- Variation in liner thickness in combination with concrete cracks can increase the liner strains even more.
- The liner strain will increase by a factor of 5-10 in comparison with the general strain level, here assumed to be the same as the maximum tensile allowable value stipulated in the ASME Code, $\varepsilon_{st} = 0.003$.
- The (safety) margin, a factor of 2-4, between the localized plastic strain levels in tension, and the strain level when rupture may occur, seems to be adequate.
- Thus, the membrane tensile steel liner strain allowable in the ASME Code seems acceptable.

1. INTRODUCTION

In Article CC-3710, liner strain allowables for Service and Factored loads are specified. Calculated liner strains, when specified loads are applied, shall not exceed the values given in Table CC-3720-1. The design calculations of the liner strains do not normally explicitly include the effect of concentration of plastic liner strains due to concrete cracking. The aim of this report is to investigate the level of tensile plastic strain reached locally due to concrete cracking, if the strain level in the liner adjacent to the cracked area has reached the maximum allowable value specified in the ASME Code (Table CC-3730-1), $\varepsilon_{st} = 0.003$. Additionally, the combined effect of concrete cracking, the friction between the liner and the concrete wall, and discontinuities such as local thinning of the liner, is discussed. The calculated maximum plastic strains at a concrete crack is then compared to the estimated strain level when liner tearing may occur.

A Ph D Thesis (Anderson, 2007) presents an analytical procedure to calculate the concentration of plastic strains due to concrete cracks in containment walls including the effect of friction between the liner and the concrete, with and without local thinning of the liner. Also, in Anderson (2007) the liner strain increase at vertical bends near an equipment hatch is studied. The work carried out by Anderson is summarized below:

It has been shown in containment scale-tests that the global displacement measured at liner failure does not correspond to the critical strain level for the liner. A general conclusion from these results is that some type of strain concentration has to take place to get this "early" liner failure. The main reasons for strain concentrations are in this paper assumed to be; (1) concrete through-wall cracking and (2) discontinuities such as penetrations. In the presented study analytic results are compared to results from a 1:4-scale containment model tested by over-pressurization in year 2000 at Sandia National laboratories (Sandia 1:4). Effects of through-wall cracks (1) are studied in a general section in mid-height of the containment. From this study it is concluded that concentrations of plastic strains could be highly increased by the friction between the liner and the concrete, especially for thin liners. The effect of discontinuities (2) is exemplified by an FE-analysis of the penetration region in Sandia 1:4 where the first liner tears appeared. From this analysis it is shown that of out-going folds in the liner tends to be straighten out when the containment expands. It is concluded that this behaviour most likely contributed to the tears found in the penetration region of Sandia 1:4.

The analytical methods presented in this paper is herein used to calculate the increase in tensile strain due to concrete cracking and thinning of the liner for a typical concrete containment wall, when it is assumed that the general tensile strain in the liner is equal to the maximum tensile strain allowable stipulated in the ASME Code (Table CC-3730-1), $\varepsilon_{st} = 0.003$. This to investigate the safety margin regarding the localized plastic strains in the liner then occurring, to the strain level when liner tearing may occur.

2. CONCENTRATION OF PLASTIC STRAINS DUE TO CONCRETE CRACKS

2.1 Introduction

When concrete cracks arises in the concrete wall to which the steel liner is attached, there may be a localization of liner strains. If it is assumed that there is no movement between the concrete and the liner at the continuous anchors, and the concrete crack spacing is larger than the spacing of the anchors, the strain in the liner field where the crack occurs will be larger than the general tensile strain in the liner. If there also exists a friction force between the liner and the concrete due to internal overpressure acting on the liner pressing it against the backing concrete wall, there will be an additional increase in tensile liner strain. A local thinning of the liner at the crack will further increase the strain. This is explained in the following sections.

2.2 Analysis Model

The analysis of the liner behaviour near a through-wall concrete crack is studied by a generalised horizontal segment at mid height of the containment wall (see Figure G-24). The liner failure mode is studied in the hope direction and effects from the vertical direction are neglected.

The parameters which are assumed to influence the liner behaviour in a generalised segment is the concrete crack width (w), the distance between liner connectors (c), the internal pressure (p), the friction coefficient between concrete and liner (μ) and of course the liner material and thickness (t).

Crack formation in reinforced concrete has a random nature, where the maximum space between cracks l can be in the order of two times the minimum crack space. The strain in the concrete between the cracks is assumed to be insignificant, i.e., all strain is assumed to be concentrated to the crack. If the general liner strain ε is known, the concrete crack width can then be calculated as $w = l \cdot \varepsilon$.

Friction between the liner and the concrete wall will be activated due to the internal pressure load (p) which presses the liner to the concrete. In this study the contact pressure is assumed to be equal to the internal pressure load (p). For containments in general, the axial stiffness of the liner is low compared to the concrete wall stiffness and therefore the contact pressure will be close to the internal pressure load (p). The friction coefficient (μ) between steel and concrete has been investigated in several research studies. From a friction test made by Anderson (2007), μ is concluded to be around 0.6.

To study the liner strain localization around a concrete crack, a one-dimensional bar model is used (see Figure G-25). The model describes a liner segment between two liner anchors, with fixed support at one end and a displacement (w) corresponding to the crack width at the other end. The friction between liner and concrete is modelled by a uniformly distributed axial load (μp). The

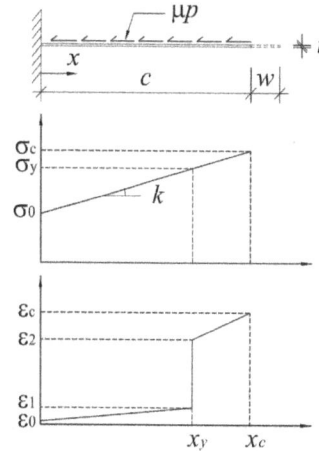

FIGURE G-25 STRESS AND STRAIN ALONG THE BAR MODEL.

steel material for the bar model is given non-linear characteristics (see Figure G-26).

The stress distribution (see Figure G-25 above) along the model will be linear with a slope (k) governed by equilibrium with the friction force.

$$k = \frac{\mu p}{t} \quad (1)$$

The stress level and also the strain distribution depend on the stress-strain relation for the steel material. When the stress in the crack section (σ_c) exceeds the yield limit (σ_y) and the stress in the fixed section (σ_0) is between 0 and σ_y the strain distribution can be described as in Figure G-25. One part will be in the elastic stage (0 to x_y) and the other part will be in the plastic stage (x_y to x_c). The sum of strain along the segment will be equal to the displacement load (w).

$$w = \int_c^0 \varepsilon(x)dx \quad (2)$$

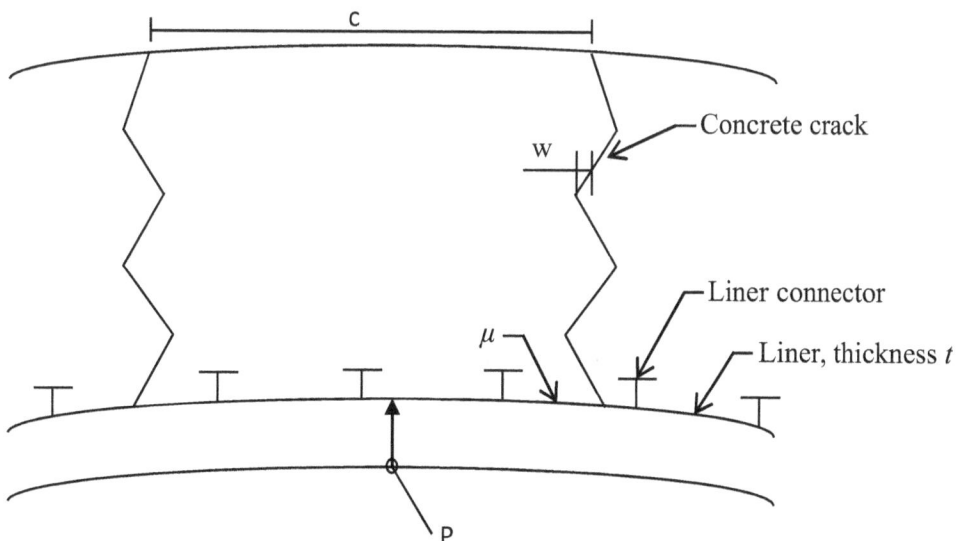

FIGURE G-24 GENERALIZED SEGMENT OF CONCRETE REACTOR CONTAINMENT WALL.

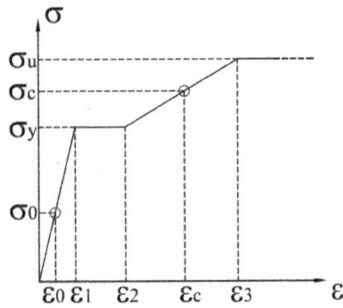

FIGURE G-26   IDEALIZED STRESS-STRAIN RELATION.

For the case $0 < \sigma_0 \leq \sigma_y$ and $\sigma_y < \sigma_c \leq \sigma_u$ an analytic expression for the σ_c can be found (see expression 3). In the expression below the simplification $\varepsilon_0 = 0$ is made. However, in the calculations presented the full expression is used.

$$\sigma_c = \frac{\varepsilon_1 E_2}{2} - \varepsilon_2 E_2 + \sigma_y$$
$$\pm \frac{\sqrt{\varepsilon_1^2 E_2^2 - 4\varepsilon_1\varepsilon_2 E_2^2 + 4\varepsilon_2^2 E_2^2 - 4\varepsilon_1 E_2 kc + 8wE_2 k}}{2} \quad (3)$$

where, $E_2 = \dfrac{\sigma_u - \sigma_y}{\varepsilon_3 - \varepsilon_2}$

2.3   Interaction between Concrete Wall and Liner

Due to the influence of the friction force between the liner and the concrete, the stress along the liner field will vary as is shown in Figure G-25, giving rise to a localization of the plastic liner strain near the crack (see Figure G-25), the strain is not evenly distributed along the liner field.

Using this model and the description of the material given in Figure G-26, it can also be concluded that the capacity of the liner is exhausted when ε_3 is reached even if the ultimate strain capacity (ε_u) is much larger (ε_u is in the order of three times ε_3). The reason for this is that when ε_3 is passed the stress in the liner will not increase and then the remaining displacement cannot be distributed along the liner due to the friction force. This means that in theory the remaining displacement will be localized to an infinite small part in the cracked section which will lead to infinite large strains i.e., failure. ε_3 then constitutes the maximum acceptable liner strain to prevent failure of the liner.

To sum up, the interaction between concrete wall and liner will have the following influence on the liner strain:

- The friction between the liner and the concrete wall will concentrate the plastic strains to through-wall concrete cracks.
- Due to the influence of friction the liner cannot be stretched beyond the hardening part. When the characteristic steel curve is flattened out after the hardening part, the remaining displacement will be concentrated to the crack location.

2.4   Thinning of Liner

A concrete crack in combination with friction is one factor causing concentrations of strains in liners as described above. Varying liner thickness could also give locally increased liner strains. To exemplify the combined effect of a concrete crack and

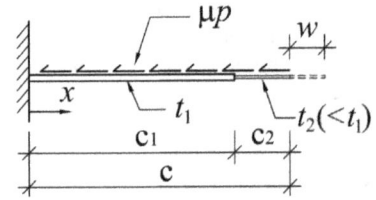

FIGURE G-27   MODIFIED BAR MODEL WITH A THINNER PART NEAR THE CRACK.

discontinuities in the form of a locally thinner liner the model used above is modified as in Figure G-27.

In Figure G-28 the stress along the liner is shown in principle for a liner with constant thickness and for a liner with varying thickness. The stress in the cracked section increases for the case with varying liner thickness in comparison with the constant thickness liner.

To sum up, variation in liner thickness in combination with concrete cracks can increase the liner strains significantly.

3.   LINER TENSILE STRAIN CALCULATIONS

3.1   Introduction

To reexamine the tension liner strain allowable given in the ASME Code (Table CC-3730-1), the maximum allowable tensile strain for factored loading, $\varepsilon_{st} = 0.003$, is applied into the analysis model described in chapter 2. Then the value of the localized strain at a concrete crack can be estimated, and compared to the maximum allowable strain, ε_3 according to Section 2.3.

3.2   Material Parameters

The following material parameters are used in the analyses:

Material parameters
$E = 220$MPa
$\varepsilon_2 = 1.3\%$
$\varepsilon_3 = 6.5\%$
$\sigma_y = 380$MPa
$\sigma_u = 500$MPa

The meaning of the parameters are given in Figure G-26.

3.3   Model Parameters

The following model parameters are used in the analyses:

Model parameters	
$w = 2$ mm	
$\mu = 0.6$	
$p = 1.0$MPa	
$t = 6.4$ mm	
Case 1: c = 450 mm	*Case 2: c = 150 mm*

The meaning of the parameters are given in Figure G-24. Case 1 ($c = 450$ mm) represents a general part of the containment wall,

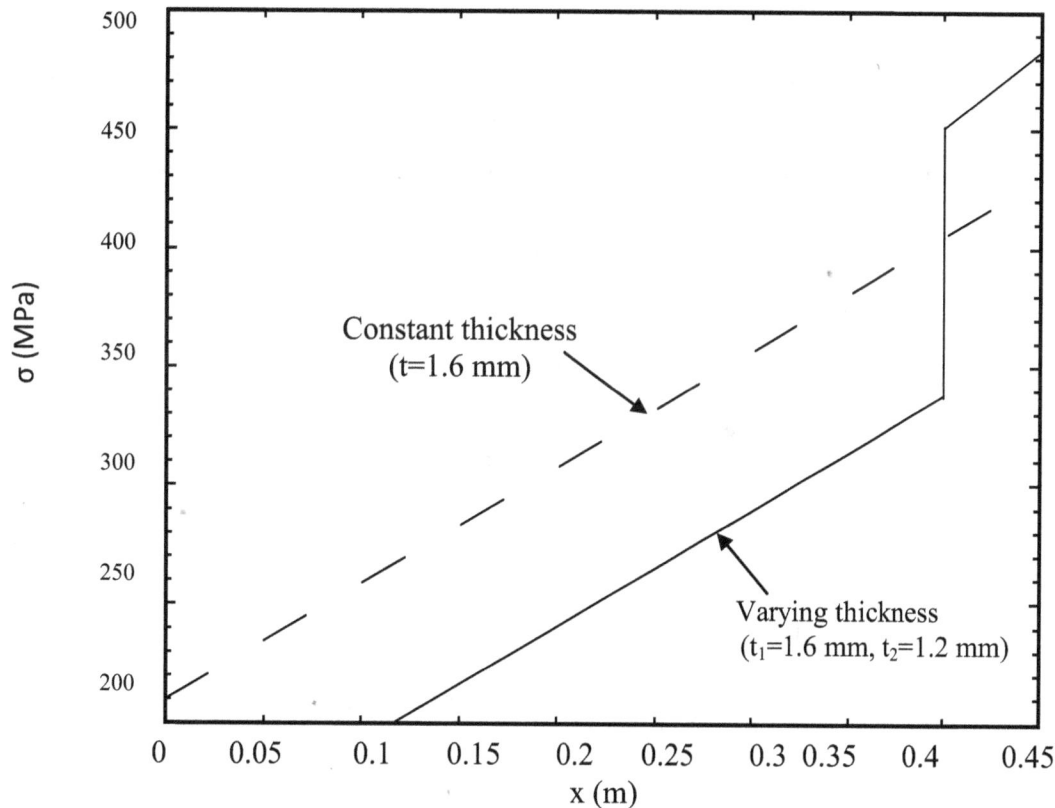

FIGURE G-28 PRINCIPLE DISTRIBUTION OF STRESS ALONG THE LINER, CONSTANT THICKNESS AND VARYING THICKNESS RESPECTIVELY.

while case 2 ($c = 150$ mm) represents a specific area of the containment wall where the anchors are placed more dense, for example near an equipment hatch.

The maximum crack space is calculated according to Eurocode 2, giving a crack spacing in the order of 600 mm for a 1 m thick containment wall. Then the concrete crack width is calculated as $w = l \cdot \varepsilon_{st} = 600 \cdot 0.003 = 2$ mm.

When studying a local thinning of the liner, it is assumed that the thickness of the liner is reduced by 10% locally near the concrete crack ($t_2/t_1 = 0.9$, see Figure G-27). The length of the thinned part, c_2, is 50 mm. The study of the effect of a thinning of the liner is denoted *Case 3*.

3.4 Analyses Results

By using the analytical method described in Chapter 2, together with the material and model parameters given in Sections 3.2 and 3.3, the localized plastic steel liner tensile strain at a concrete crack in the containment wall has been calculated under the assumptions that the general strain in the liner is the same as the maximum allowable membrane tensile strain specified in the ASME Code, Table CC-3730-1, $\varepsilon_{st} = 0.003$.

This analysis is used to compare the localized plastic strain value achieved when the liner is in a general tensile strain state equal to the maximum value allowed in the ASME Code, with the strain level when liner rupture may occur ($\varepsilon_3 = 6.5\%$ in Figure G-26, as explained in Section 2.2), to estimate the safety margins to rupture in the liner when utilizing the ASME Code tensile membrane strain allowable.

Summary of analyses results:

Analyses results				
Case	General strain (%)	Max. stress (MPa)	Localized strain at concrete crack (%)	Strain level when rupture may occur, $\varepsilon_3(\%)$
Case 1: Anchor spacing 450 mm	0.3	389	1.7	6.5
Case 2: Anchor spacing 150 mm	0.3	391	1.9	6.5
Case 3: 10% local thinning	0.3	412	2.9	6.5

It is seen from the results that the general tensile liner strain is increased by a factor of 5-10 due to the localization of the strains that occurs at concrete cracks. However, the safety margin to the strain level when rupture may occur is still acceptable, a factor of safety somewhere between 2-4.

4. CONCLUSIONS

From this study, the following conclusions can be drawn:

– Cracking of the concrete containment wall will give rise to localization of the tensile liner strain.

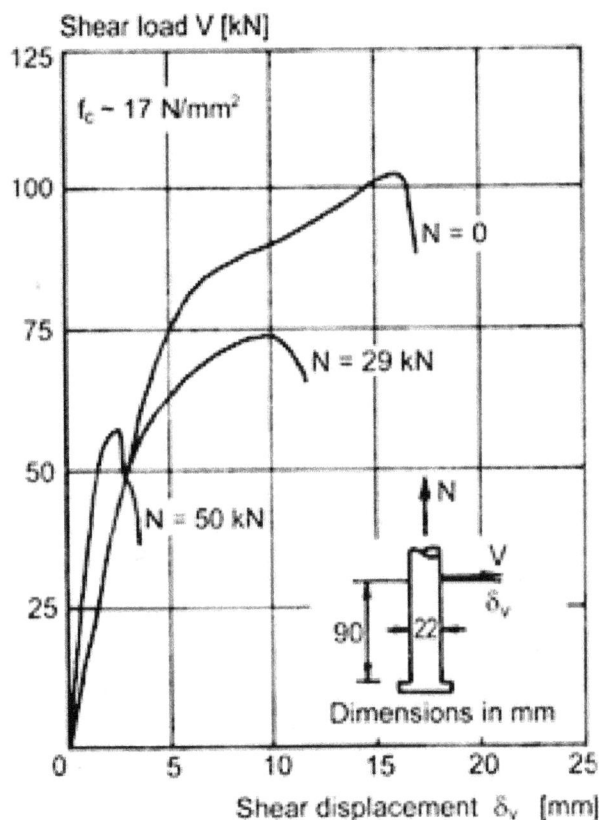

FIGURE G-29 SHEAR LOAD AS A FUNCTION OF SHEAR DISPLACEMENT FOR HEADED BOLTS SUBJECTED SIMULTANEOUSLY TO VARIOUS LEVELS OF TENSION LOADING (AFTER BODE (1985), TAKEN FROM ELIGEHAUSEN (2006)).

- The friction between the liner and the concrete wall will concentrate the plastic strains to through-wall concrete cracks, giving a further increase of the localized strain level.
- Due to the influence of friction the liner cannot be stretched beyond the hardening part. When the characteristic steel curve is flattened out after the hardening part, the remaining displacement will be concentrated to the crack location.
- Variation in liner thickness in combination with concrete cracks can increase the liner strains even more.
- The liner strain will increase by a factor of 5–10 in comparison with the general strain level, here assumed to be the same as the maximum tensile allowable value stipulated in the ASME Code, $\varepsilon_{st} = 0.003$.
- The (safety) margin, a factor of 2-4, between the localized plastic strain levels in tension, and the strain level when rupture may occur, seems to be adequate.
- Thus, the membrane tensile steel liner strain allowable in the ASME Code seems acceptable.

5. REFERENCES

Anderson P. (2007) Structural integrity of prestressed nuclear reactor containments, PhD Thesis, Lund university.

Anderson P. (2008) Concentration of plastic strains in steel liners due to concrete cracks in the containment wall. Int J Pressure Vessels Piping, doi:10.1016/j.ijpvp.2008.03.004.*

Anderson P. (2008) Analytic study of the steel liner near the equipment hatch in a 1:4 scale containment model. Nucl Eng Des, 238: 1641–1650.*

Anderson P., Jovall O. (2007) Increased plastic strains in containment steel liners due to concrete cracking and discontinuities in the containment structure. Paper 1490, 19th International Conference on Structural Mechanics in Reactor Technology, Toronto 2007.

[G-4] LINER ANCHORS INTERACTION OF TENSILE AND SHEAR FORCES, O. JOVALL, SCANSCOT TECHNOLOGY AB, MARCH 2008

For the design to be in accordance with CC-3123 (a) stating that the liner anchor system shall be designed to accommodate all design loads and deformations without loss of structural or leak-tight integrity, in addition to the fulfillment of the liner anchor allowables stated in Table CC-3730-1, the resistance to combined tensile and shear loads shall be considered in design.

In Figure G-29 the effect of simultaneously acting tensile and shear forces for a punctual anchor (headed stud) is exemplified. From Figure G-29 it is seen that both the shear force capacity and

---

* Papers concerning increased liner strain included in Ph D Thesis, Anderson (2007).

the shear displacement capacity is reduced significantly when a tensile force is acting on the anchor. For example, for a tensile load of 50 kN, the shear load capacity is reduced from 103 kN to 58 kN (almost 45% reduction of the capacity), and the shear displacement capacity is reduced from 17 mm to 3 mm (more than 80% reduction of the capacity).

It is obvious that the combined loading shall be considered for mechanical loads if both a tensile force and shear force is acting,

but also for displacement limited when simultaneously acting tensile and shear forces are present as shown in Figures G-30 and G-31. This combined effect for displacement limited loads is normally not of a major concern for continuous anchors (due to the fact that often $N_{ua} \leq 0.2N_n$, see below), however it may be for punctual anchors.

Resistance to combined tensile and shear loads shall be considered in design using an interaction expression that results in computation of strength in substantial agreement with results of

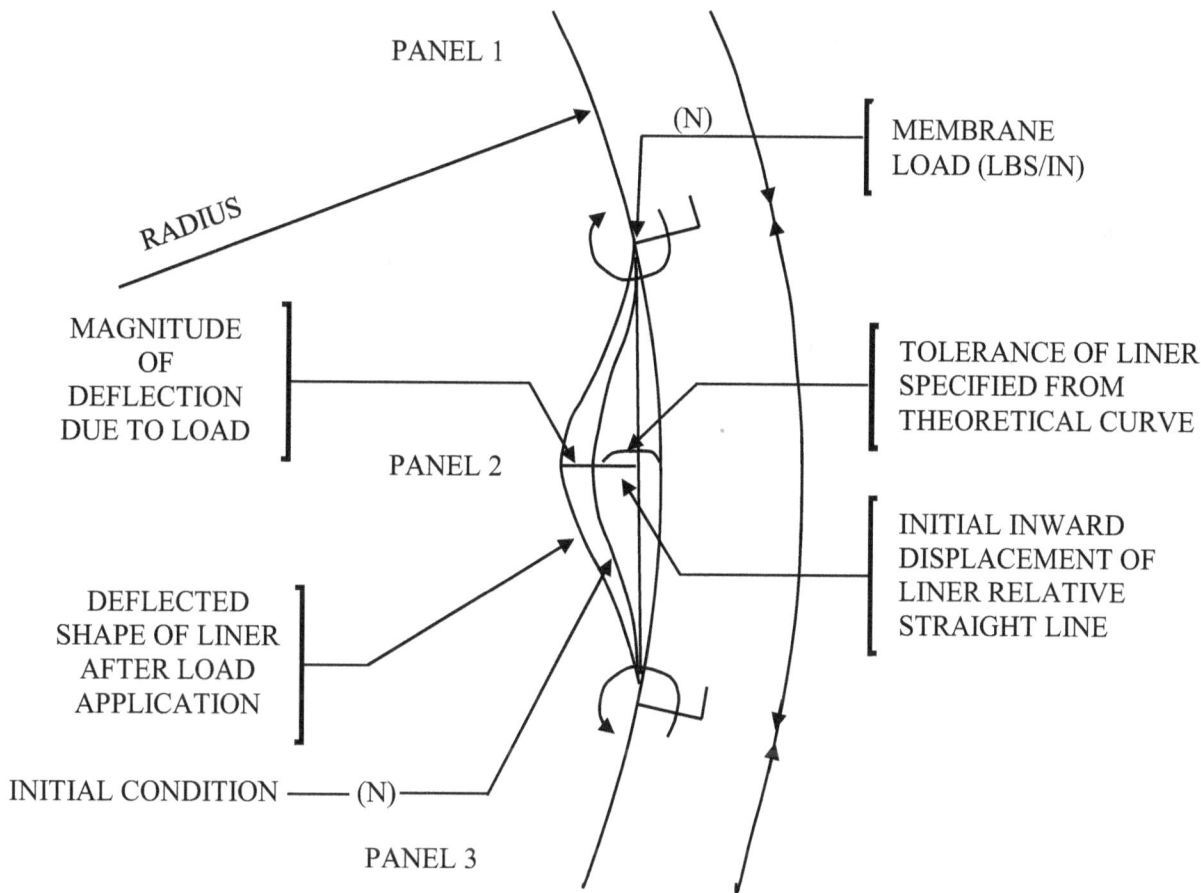

FIGURE G-30   DISPLACEMENT LIMITED IN-PLANE SHEAR LOADS ACTING ON LINER ANCHOR (BECHTEL CORPORATION (1972)).

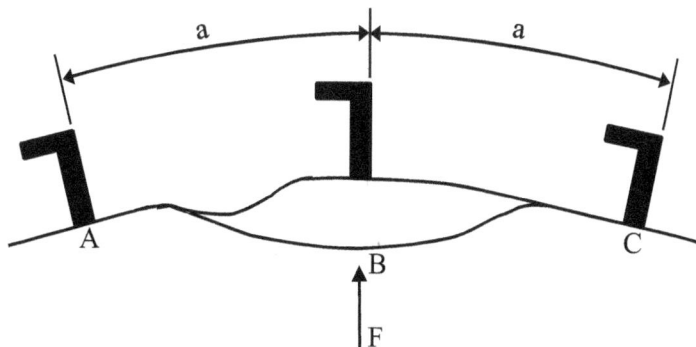

FIGURE G-31   DISPLACEMENT LIMITED TENSION LOADS ACTING ON LINER ANCHOR (BECHTEL CORPORATION (1972)).

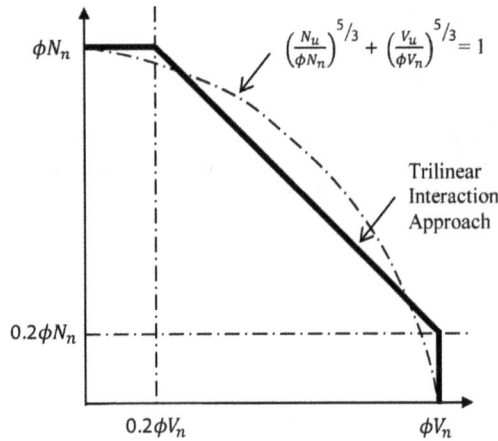

FIGURE G-32 SHEAR AND TENSILE INTERACTION EQUATION (ACI-349 (2007)).

comprehensive tests. This requirement shall be considered satisfied by the following, where V_{ua} and N_{ua} is the simultaneously acting shear and tensile force respectively, V_n the shear force strength, and N_n the tensile force strength:

If $V_{ua} \leq 0.2\, V_n$, then full strength in tension shall be permitted: $N_n \geq N_{ua}$.

If $N_n \leq 0.2\, N_{ua}$, then full strength in shear shall be permitted: $V_{ua} \geq V_n$.

If $V_{ua} > 0.2\, V_n$ and $N_{ua} > 0.2\, N_n$, then $(N_{ua}/N_n) + (V_{ua}/V_n) \leq 1.2$.

Any other interaction expression that is verified by test data, however, can be used to satisfy the requirements. This includes the possibility to by testing determine F_u or δ_u when the second force is present.

The shear-tension interaction expression has traditionally been expressed as $(N_{ua}/N_n)\xi + (V_{ua}/V_n)\xi \leq 1.0$ where ξ varies from 1 to 2. The current trilinear recommendation is a simplification of the expression where $\xi = 5/3$ (Figure G-32). The limits were chosen to eliminate the requirement for computation of interaction effects where very small values of the second force are present.

REFERENCES

ACI, Code Requirements for Nuclear Safety Related Concrete Structures (ACI 349-06) First Printing September 2007.

Bechtel Corporation, Containment Building Liner Plate Design report, BC-TOP-1, Rev. 1, December 1972.

Bode H., Hanenkamp W., Zur Tragfähigkeit von Kopfbolzen bei Zugbeanspruchung (Load-bearing capacity of headed anchors under tension loads), Bauingenieur 1985, pp. 361–367 (in German).

Eligehausen R. et al., Anchorage in Concrete Construction, Ernst & Sohn Verlag, 2006.